CONFLICT IN HISTORY, MEASURING SYMMETRY, THERMODYNAMIC MODELING AND OTHER WORK

Dennis Glenn Collins

authorHOUSE®

AuthorHouse™
1663 Liberty Drive
Bloomington, IN 47403
www.authorhouse.com
Phone: 1-800-839-8640

First published by AuthorHouse   11/08/2011

ISBN: 978-1-4670-7641-8 (sc)
ISBN: 978-1-4670-7640-1 (ebk)

Library of Congress Control Number: 2011919897

Printed in the United States of America

Conflict in History, Measuring Symmetry, Thermodynamic Modeling and Other Work by Dennis Glenn Collins

TABLE OF CONTENTS

Remark: by Author:
The work is far from being self-contained; even consulting the references, there are gaps due to not being able at present to retrieve work prior to Microsoft Word. This restriction means none of the listed thermodynamic modeling papers are included, although some can be obtained from other publications. Nonetheless the author thanks AuthorHouse for work on this book. Nov. 1, 2011

LIST OF ILLUSTRATIONS

Conflict in History Study of the Mexican (-American) War 1846-1848

By Dennis Glenn Collins
7108 Grand Blvd.
Hobart, IN 46342-6628
Jan. 10, 2020

This study of the Mexican (-American) War 1846-1848 is based on the author's 1971 paper "Conflict in History," whereby events supposedly favorable to one side (in this case the United States) are charted above the horizontal axis and events apparently favorable to the other side (Mexico) are graphed below the horizontal or time axis, which in this case is divided into years. Please see Figure 1.

The dates of deaths of U.S. Presidents and past presidents are taken to calibrate the study, namely at the plus and minus two periods, the death of Andrew Jackson June 8, 1845 and James Polk June 15, 1849, both from Tennessee and symbolic of American expansion Westward. Additionally at points favorable to Mexico are the deaths of Pres. Zachary Taylor July 9, 1850 and John Quincy Adams Feb. 23, 1848. This pattern continued in U.S. dealings with Latin America with the 1901 assassination of McKinley after the Spanish-American War of 1898 and John Kennedy in 1963 after the 1962 Bay of Pigs invasion. Teddy Roosevelt died after an expedition into the Amazon.

The decision point therefore comes out about June 11, 1847, favorable to the U.S. The successful campaign of Gen. Winfield Scott into Mexico from March 9, 1 847 (Veracruz) to Sept. 14, 1847 (Mexico City) three months before and after June 11, 1847 more or less bracket this decision point, following up a victory in at Buena Vista Feb. 22-23, 1847 by Zachary Taylor in northern Mexico. Also near the decision point July 1847 the Mormons settled around Salt Lake City, Utah. Thus the decision point can be taken as the success of the "Manifest Destiny" of the U.S. to win the West.

The supposed one-year advance of Mexican interests took place from December 1845 to December 1846. This year saw Mexico with a certain moral advantage since it was understood the entrance of Texas into the Union, as happened Dec. 29, 1845, would mean war with the U.S. The actual war was declared by Congress May 13, 1846 after the so-called Thornton affair saw 16 Americans die after a clash April 25, 1846. In spite of Mexican defeats at the hand of Zachary Taylor at Palo Alto May 8 and Resaca de la Palma May 9, 1846, Mexico under Santa Anna seems to have recovered sufficiently to outnumber Zachary Taylor's troops 4 to 1 at the above-mentioned battle of Buena Vista, Feb. 22-23, 1847. There was also opposition in the U.S. Congress to the Mexican War, which was seen as a scheme to increase slave-holding territory.

Along these lines, Abraham Lincoln's Dec. 22, 1847 "Spot" Resolutions against Pres. Polk were supposedly at a period favorable to Mexico. However these were dropped and the period Dec. 1847 to Dec. 1848 saw a one-year advance of U.S. "Manifest Destiny" interests, ending with the election of Zachary Taylor as President in

Nov. 1848, even though he hadn't bothered to vote. The one-year advance saw the discovery of gold in California, Jan. 24, 1848 (leading to the Gold Rush), and the addition of the territory, including Texas and California, sought by Pres. Polk at the Treaty of Gaudalupe-Hidalgo Feb. 2, 1848 under the questionable direction of Nicholas Trist. Actually this treaty could be considered as under the period favorable to Mexico, since some Americans wanted to annex all of Mexico, an interesting possibility considering present difficulties. Also as part of the one-year advance the Oregon dispute with Britain was mostly settled.

At a point favorable to the U.S., Nov. 1844, James Polk was elected Pres. of the United States. Further topics for study include supposedly favorable events Dec. 1849 and events at the so-called "future" points, approximately Feb. 1846 and Oct. 1847.

Remark: Also the supposed "high points" of the Mexican side could be studied more. These Mexican high points are somewhat limited because at the 25-years cycle, the U.S. was at a high (1848) in between the lows of the Alamo (1836) and the Civil War (1861). Consequently the Mexican high points are significantly attenuated.

Mayaguez, Puerto Rico Jan 10, 2010.
Dennis G. Collins

REFERENCES

Bailey, Thomas, David Kennedy and Lizabeth Cohen, *The American Pageant*, Houghton—Mifflin, Boston 1998
Wikipedia article on "Spot Resolution"
Wikipedia article on "Mexican-American War"
World Almanac and Book of Facts (2003 WRC media) articles on Presidents and States
World Book (1973) article on "Mexican War"

BIN LADEN STUDY

By Dennis G. Collins
1519 S. State Rd 119 Apt. 2
Winamac, IN 46996-8550
Sept. 24, 2011

According to the author's 1971 theory "Conflict in History," the chart plots events involving the death of Al Qaeda leader Bin Laden on May 1, 2011, with events favorable to the U.S. plotted above the time-coordinate axis and events favorable to Al Qaeda plotted below the t-axis in 27-day periods. The decision point around June 24, 2011 favorable to Al Qaeda appears to be the U.S. move to draw down troops in Afghanistan, announced by Obama at a "future point" July 4, 2011. The U.S. high points at the +/-2 period points are the deaths of Bin Laden May 1and supposed #2 man al-Rahman Aug. 22. A U.S. low point was the shooting down of a Navy Seal team Aug. 5, 2011 at the +1.5 period point.

CONFLICT IN HISTORY
STUDY OF WORLD WAR I

By Dennis G. Collins, 7108 Grand Blvd., Hobart, IN 46342-6628
Urb. Mayaguez Terrace, 6009 Calle R. Martinez Torres,
Mayaguez, PR 00682-6630

This analysis of World War I (Please see Figure 1.) is based on the author's 1971 paper "Conflict in History," which graphs "warp" above the time axis as favorable to one side, and "warp" below the axis as favorable to the other sides in a conflict, according to the formula: $f(x)=\exp(-(.75*t)^2/2)*(\cos(2*Pi*t)*(4*(.75*t)^2-2))$. Here events favorable to the Allies (Britain, France, Russia, later the United States (April 6, 1917), Italy and others) supposedly come above the time axis and events favorable to the Central Powers (Germany, Austro-Hungary, later the Ottoman Empire—Turkey— and others) supposedly come below the time axis. More generally, the war can be considered as a conflict between democracy (top) and empire (bottom). With his "war to end wars," Woodrow Wilson played the role of theoretician, entering stage left as U.S. President in Nov. 1912 and more or less exiting stage right with a stroke about Oct. 2, 1919. In this role he corresponds to Winston Churchill entering May 1940 in World War II.

Amazingly, both World War I and World War II (previously studied, Please see Figure 2.) follow the same pattern, with World War II events following World War I events by approximately 27 years and 8 months, i.e. about one generation. Although the graph patterns are exactly the same, the actual events are mostly "variations on a theme," as evidenced by the fact that no historians have noticed this fact so far as known to the author. The correspondence may be set up by the similar role of Winston Churchill in the failed Gallipoli campaign of World War I starting April 25, 1915 and his apparent role in the failed Dieppe raid along the French coast of August 19, 1942, which both came during the "one-year advance" of the opposition.

The decision point of the war is taken as April 1916, with the Battle of Verdun starting Feb. 21, 1916 halting the German advance with Petain's cry "They shall not pass," and going until June 1916, and the naval Battle of Jutland, May 31-June1, 1916, which, apart from submarine warfare, gave Britain control of seapower. Verdun, east of Paris and close to the triple French/Belgian/German border, played the role of hinge of a door swinging up and down along the English Channel coast. As long as the French controlled this hinge Verdun, it was unlikely the Germans could push it open sufficiently to take Paris. Along the "door" to the coast were the trench lines that moved back and forth slightly with hundreds of thousands of casualties as each side pushed on the door.

Here April (4 months) 1916 plus 27 years and 8 months= 12 months and 1943= Dec 1943, the decisive month of World War II according to the previous study.

The word "one-year advance" is not entirely accurate since the best way to "advance" was frequently to maintain a defensive posture and mow down the enemy with machine gun fire as it tried to take more territory. When the war started shortly after the assassination of Archduke Francis Ferdinand June 28, 1914, the Germans swept through Belgium and threatened to push open the door sufficiently to take Paris. The Allies stopped them in the 1st Battle of the Marne, Sept. 6-9, 1914 in a high point for the Allies. However there followed a one-year "advance" of the Central Powers as they kept control of industrialized northern France and most of Belgium, while employing submarine

warfare, including the sinking of the U.S. oceanliner *Lusitania*, May 7, 1915 and poison gas for the first time April 22, 1915. Also Germany expanded eastward ("Drang Nach Osten") after crushing the Second Russian Army at Tannenburg, Aug. 26-31, 1914. The Central Powers' one-year advance included decimating the above-mentioned Gallipoli campaign April 25, 1915 to about December 1915, forcing Winston Churchill to resign from the Admiralty.

The "future points" approximately 4-5 months before and after the decision point April 1916 saw the development of general relativity in Germany by Einstein and Hilbert in Berlin, Germany in Dec. 1915 and the first use of tanks by the British Sept. 15, 1916. These events foreshadow to some extent what the author calls the "Neo-Gorgon" religion, including the ability of mass to deflect light.

The supposed one-year advance of the Allies from Oct. 1916 to Oct. 1917 began with the death of aged Austrian Emperor Franz Joseph Nov.21, 1916, since he was not readily replaced. The German push to the east was compatible with the victory of democracy over empire since the Russian system was more despotic than the German. Thus the one-year advance of democracy ended with the abdication of the czar and the short-lived Kerensky democracy (being a high-point for democracy), and a return to despotic power Nov. 7, 1917 with the Bolshevik Revolution.

In the west, German declaration of unrestricted submarine warfare Feb. 1, 1917caused the United States to enter the war on the Allies' side April 6, 1917. However the Allies attempt to push the trench warfare door to the north in the 3rd Battle of Ypres July 31-Nov.10, 1917 was stopped by the Germans at the end of the period of one-year advance (Oct.17, 1917), and the pendulum swung back to the German side. In a high point for the Central Powers (about April, 1918), the Germans were able to switch resources from the recently-victorious eastern front and mount 4 offensives trying to push the door back south. As usual none of these offensives (March 21, 1918—2nd Battle of Somme, April 9, Ypres, May 27, Aisne code-named Blucher, July 15 on—2nd Battle of Marne) were successful, and German forces were depleted.

Meanwhile the entry of America into the war gradually swung the pendulum (or door) back to the Allies' side (north), and American forces, for example at Chateau-Thierry, July 21, 1918, were able to spearhead the final Allied offensive, leading to the high-point Oct. 1918 collapse of German forces. Shortly thereafter the empires of the Central Powers were gone with Armistices (Sept. 29, 1918 Bulgaria, Oct. 30, Ottoman Empire, Nov. 3 Austria, and 11 a.m. Nov. 11, 1918 Germany).

A Woodrow Wilson attempt at a lenient peace based on his 14 points of Jan. 1918 failed and the harsh Treaty of Versailles after July 1919 was another high point for the Allies versus the Central Powers.

Among events that seem not to fit the theory are the surrender of the British at Al-Kut (Kut-al-Amara) to the Turks on April 29, 1916, although the eventual British occupation of Jerusalem Dec. 9, 1917 does occur at a supposed high point for the Allies. Wounding of 304th Tank Brigade commander Lt. Col. George S. Patton late Sept. 1918 in the Allied Meuse-Argonne offensive is somewhat out of phase with his death Dec. 21, 1945 in World War II. Also the commissioning of the 5 "Kaiser" class German battleships 1912 (Aug 1, Aug 1, Oct 15) and 1913 (May 14, July 31) are more in phase with the German sweep through Belgium than the 6 months earlier that the theory might predict.

REFERENCES

World War I, *The World Book Encyclopedia*, Chicago, IL 1974.
Macropedia article on The World Wars, *The New Encyclopedia Britannica*, Chicago, IL 1990.
Mayaguez, Puerto Rico Jan. 6, 2009

THE WAR OF 1812

By Dennis Glenn Collins
1519 S. State Rd. 119, Apt. 2
Winamac, IN 46996-8550
October 8, 2011

This paper studies the War of 1812 according to the author's 1971 theory "Conflict in History. The time periods favorable to the United States are plotted above the horizontal or time axis t and the periods most favorable to the British are plotted below the horizontal axis, measured in years. Consistent with the alternative name of the war, "Mr. Madison's War," U.S. President James Madison enters stage left in Jan 1809 after being elected in Nov. 1808 and taking office in March 1809; he leaves stage right in Jan. 1817, leaving office in March 1817. Please see Figure 1.

Since the War of 1812 was in many respects a backwater of the war against Napoleon by the British, the decision point is taken as Jan. 1813 after Napoleon's disastrous invasion of Russia in the autumn of 1812, with the last French troops leaving Russia Dec. 14, 1812. Other British "high" points (at the bottom of the chart) were the abdication of Napoleon April 6, 1814 and his exile to Alba, and next year his defeat at the Battle of Waterloo, June 18, 1815.

The war in America was mainly a defensive struggle, with major offensives leading to reversals. In football terms one could say scoring drives ended in interceptions or fumbles. The roots of the war came from British or Jewish interference in U.S. affairs, including impressment of U.S. sailors to help the war against Napoleon, and foreign influence in the First Bank of the United States, whose charter expired in 1811. However U.S. difficulties funding the war led to a 2nd Bank of the United States being created and signed into law in April 1816, with subscriptions starting at a British "high" point July 1816. As with the Treaty of Ghent Dec. 24, 1814, which supposedly returned things to the status quo, the 2nd U.S. bank opening Jan. 7, 1817 in many ways returned financial control to rich/Jewish or European elites until the 1830's, apparently after the substantial Jewish takeover of British finance after Warterloo due to better communications.

The supposed one-year advance of the U.S. interests from July 1811 to July 1812 saw failure of diplomacy to resolve issues with Britain, leading to declaration of war against Britain by the U.S. Congress June 18, 1812 at the instigation of "War Hawks." A British attempt to avert war by cancelling some orders of Council allowing impressment came too late to reach U.S. shores, which communication problems seemed to happen repeatedly in the war. Supporting the War Hawks, the U.S. defensive victory under William Henry Harrison at the Battle of Tippecanoe Nov. 7, 1811 in Indiana is called for practical purposes the opening battle of the War of 1812. The one-year advance saw a three-prong attack against British Canada, with each prong failing as time went beyond the one-year advance. The result was British capture of American armies and victories at Queenston Heights Oct. 11, 1812 and Frenchtown, Jan 19, 1813. With diminished threat from Napoleon after April 6, 1814, the British launched their own three-prong attack (Champlain, Chesapeake, and New Orleans), again each prong of which failed, after going beyond the one-year period of advance. The Battle of Lake Champlain/Plattsburg Sept. 11, 1814 turned back the Champlain prong; the Battle of Bladensburg and burning

of Washington, D.C. buildings August 24, 1814 (U.S. forces had burned some of York/Toronto April 1813) followed by the death of the British leader Ross Sept 12, 1814 and stalling of the British offensive in Baltimore Sept. 13-14, with the composition of the Stars and Stripes national anthem by Francis Scott Key, turned back the Chesapeake prong; and the Battle of New Orleans U.S. victory Jan. 8. 1815 under Andrew Jackson and with the death of the British commander Pakenham, ended the New Orleans prong.

Naval U.S. high points were the defeat of the British ship *Guerrierre* by the *U.S. Constitution* ("Old Ironsides") August 9, 1812 at one of the "future points," and the Battle of Lake Erie victory of Oliver Hazard Perry Sept. 9-10, 1812, which allowed the recapture of Detroit. The Thames River Battle Oct. 5, 1813 saw the Indian leader Tecumseh die, pretty much ending the British-Indian alliance. Another British loss during U.S. high points was the assassination of British Prime Minister Spencer Percival May 11, 1812, partly leading to the war. Also Major-General Sir Isaac Brock, responsible for Canadian defense, died Oct. 13, 1812 at the Battle of Queenston.

The war saw little change along the U.S./Canadian border, although the War Hawks gained some goals more-or-less indirectly with the departure of Indians, French and British from the U.S. Midwest, leading to the admission of Indiana to the Union Dec.11, 1816, followed by other Midwest and Southern states. Britian gained some by access to financing Western expansion.

REFERENCES

"The War of 1812" article in *The World Book* encyclopedia, Vol. 19, 1960.
Atlas of American Military History, Ed. By James C. Bradford, Oxford University Press. 2003, pp 36-41.
Internet articles on Spencer Percival and Andrew Jackson and others and War of 1812.
http://americanhistory.about.com/war of 1812/a/war-of-1812-timeline.htm scatteredremanat. org/AndrewJackson.pdf http://en.wikipedia.org/wiki/Spencer_Percival http://en.wikipedia.org/wiki/Isaac_Brock articles on Napoleon

CONFLICT IN HISTORY STUDY OF 2008 U.S. PRESIDENTIAL ELECTION

By Dennis G. Collins, 7108 Grand Blvd., Hobart, IN 46342-6628
Urb. Mayaguez Terrace, 6009 Calle R. Martinez Torres, Mayaguez, PR 00682-6630
Aug. 6, 2009

This analysis of the 2008 U.S. Presidential Election (Please see Figure I.) is based on the author's 1971 paper "Conflict in History," which graphs "warp" above the time axis as favorable to one side and "warp" below the axis as favorable to the other side in a conflict, according to the formula $f(x) = \exp(-(.75*t)^2/2)*(\cos(2*Pi*t)*(4*(.75*t)^2-2))$. Here events favorable to Barack Obama are plotted above the time axis and events favorable to John McCain are plotted below the time axis, with the time scale plotted in periods of 27 days.

The decision point is taken as Sept. 24, 2008, compatible with the Internet article "Sept. 24, 2008—The Day John McCain lost the election" by Daniel Gross, apparently posted on *Slate*, Nov.4, 2008, the day of Barack Obama's win in the 2008 U.S. Presidential Election at the 1 and ½ period point (Nov.4,2008) also favorable to Obama. Sept 24 was the date of a dismal interview of Republican Vice-Presidential nominee Alaska Gov. Sarah Palin with TV anchorwoman Katie Couric as well as a speech by Pres. Bush urging passage of a $700 billion bailout package for business.

McCain's attempt to avoid the first Presidential debate Sept.26 by suspending his campaign and flying to Washington, D.C. to deal with the global financial meltdown had come up short and most gave Obama the win in the first debate after McCain returned to Oxford, Mississippi for the debate. It was felt McCain needed a "knockout" in the debate to reverse an expected Democratic victory because of U.S. economic problems (It's the economy, stupid.)

The one-period advance of the McCain campaign Aug 14, 2008 to Sept. 10, 2008 according to the theory, following his Presidential appearance in getting Russia to back off its invasion of Georgia in Asia and selection of Sarah Palin as Vice-Presidential nominee Aug. 29 and nomination as Republican Presidential nominee Sept. 4. Sept. 3, 2008 the black Detroit mayor Kilpatrick had been forced to agree to resign. However beyond Sept. 10, things began to unravel, with Palin's interview with TV anchor Charlie Gibson on Sept. 11 and the Lehmann Bros. bankruptcy Sept. 15 with McCain saying the country's "economic fundamentals are strong." Sept. 12 Hurricane Ike hit Galveston, TX contributing to the turnaround of McCain's fortunes.

The one-period advance of Obama's campaign Oct.7-Nov 4, 2008 started with Obama's respectable showing in the 2nd TV debate Oct. 7. O.J. Simpson's conviction Oct 4, 2008 had raised the possibility of a "Bradley effect" vote against Obama; however the tables were turned Oct. 27 when Alaska Sen. Ted Stevens was convicted on corruption charges (later reversed), pointing to the possibility of Palin having to resign (which has now occurred, although maybe not by necessity). A false claim of black racism by a white woman A. Todd Oct. 22 in Pittsburgh added to McCain's problems, together with no surprise knock-out in the 3rd Presidential debate Oct. 15. Early voting and an Obama TV "infomercial" Oct. 29 added to Obama's momentum. Much replayed "Saturday Night Live" TV skits against Palin after the Sept. 24 interview and against McCain Nov.1 by tying him to Bush's record came at Obama high points.

In general McCain's claim of being a steady hand to prevent nuclear war proved incompatible with the "maverick" status he adopted to distance himself from the Bush record and select Palin as VP nominee.

Another high point for Obama proved to be his selection of Hillary Clinton as Sec. of State designate Dec. 1 according to the "Team of Rivals" idea. Low points for Obama included the Nov. 19 Al Qaida video attack against him and the Blagojevich Governor scandal in Illinois Dec. 10 as well as the Madoff N.Y. scandal Dec. 11.

REFERENCES

Internet articles, includin

Conflict in History Study of the U.S. Civil War

By Dennis Glenn Collins
7108 Grand Blvd.
Hobart, IN 46342-6628
August 3, 2009

This study of the U.S. Civil War 1861-1865 is based on the author's 1971 paper "Conflict in History," whereby events supposedly favorable to one side (in this case the North or Union) are charted above the horizontal axis and events supposedly favorable to the other side (the South) are charted below the horizontal axis, which is for this study divided into years. Please see Figure 1.

The decision point at the middle of the horizontal axis is taken as July 4, 1863 marking the Union victory at Vicksburg, MS giving the Union control of the Mississippi River and the Union victory over Robt. E. Lee at the Battle of Gettysburg, PA July 1-3, 1863, the "high tide of the Confederacy "(South).

The "two-period points" before and after the decision point supposedly favorable to the South occurred about July 1861 and 1865, roughly the fall of Fort Sumpter April 1861 to the First Battle of Bull Run July 21, 1861 and the assassination of Abraham Lincoln April 14, 1865 followed by uncertainty over treatment of the South after the War.

The "one-period long" advance of the South took place roughly Jan. 1862 to Jan. 1863 and saw the Lincoln appoint three different military commanders McClellan, Burnside and Hooker who either suffered defeats or couldn't follow up on victories. McClellan re-built the Union Army of the Potomac after the First Battle of Bull Run, but displayed no offense.

The "one-period long" advance of the North under U.S. Grant took place approximately Jan. 1864 to Jan. 1865 and saw the end of threats to the Capitol with "Sheridan's ride" leading to the defeat of Jubal Early, Oct. 19, 1864 and Lincoln's re-election Nov. 1864 as well as Sherman's "March to the Sea" after taking Atlanta Sept. 3, 1864. This period may be considered as extending to April 9, 1865 with Lee's surrender to Grant at Appomattox, VA.

Other high points for the Republican/Abolitionist cause (i.e. North) included John Brown's raid on Harper's Ferry Oct. 16, 1859, Lincoln's election Nov. 1860 and the rejection of Senators and representatives from the South by Congress Dec. 1865.

It appears that to prevent the survival of the Confederacy after its one-year advance, Lincoln turned to the Emancipation Proclamation Jan 1, 1863 and an end of the previous "kid's glove" policy to get the South to rejoin the Union. This change gave a "new birth" to the Union more along lines envisioned by John Brown, with destruction of the slave-based culture of the South.

The "future points" saw the first fighting by black troups for the Union March 1863 and the Gettysburg address by Lincoln Nov. 19,1863.

REFERENCES

World Book articles on the Civil War and other topics
McPherson, James M., *Tried by War*, Penguin Press, NY 2008.

THREE FOOTHILL-VERSUS-MOUNTAINS CONFLICT-IN-HISTORY CASE STUDIES

By Dennis Glenn Collins
Urb. Mayaguez Terrace
6009 Calle R. Martinez Torres
Mayaguez, PR 00682-6630
April 6, 2010

ABSTRACT

This paper treats 3 case studies of the author's 1971 theory "Conflict in History," namely 1) the American Revolution, 2) the Theodore Roosevelt U.S. Presidency, and 3) the Viet Nam War by a 2-or-3-impulse method. A goal is to see how two impulses may fit together, toward an "interpenetrating vortex street" model. There seems to be a possibility (cases 1) and 3) above) of two impulses creating a "biased platform" on which a third higher-frequency (month versus year) impulse may act. This result is an emergent phenomenon of multi-impulse models. The three cases show very different graphs, so that a typical reader, as the typical history-book reader, would not see the underlying structure.

INTRODUCTION

My son asks if the author's "Conflict in History" theory of 1971 is falsifiable or refutable. Technically the answer appears to be "No" in the sense that continuous linear conbinations and translations of the basic impulse response can represent any function according to the theorems of functional analysis. However the practical value of the theory would rest on getting good results with only a limited number of impulses.

As far as the size of the impulses, a "Bayesian" view would take the amplitude of the impulses versus time to be approximately equal, unless there is strong reason to do otherwise. That is, the tendency of history books is to spend about the same number of pages on the same length of time period. In the case of foothills versus mountains, there IS strong reason to consider the mountains to have higher amplitude than the foothills, since, well, they do have larger amplitude versus sea level. Nonetheless in this study, the amplitudes of the impulses are taken about equal, leading to more of a history-book treatment. In the case of the American Revolution, the one-impulse (mountain) model has already been presented. This model is compared to a 3-impulse model. In the other two cases, the mountain (1-impulse) model is presented and compared to a 3-impulse model.

To review the model, the events supposedly favorable to one side of the conflict are graphed above the horizontal time axis, and events favorable to the other side are graphed below the horizontal axis. The maximum amplitude points occur plus or minus two periods (time units) before and after and on the opposite side of a central impulse or decision point. There are 1-year periods of "momentum" favorable to one side or the other from ½ to 1 ½ periods before and after the central decision point. There are "future" points about .4 time units before and after the central decision point. The impulse pattern is the same for every conflict; however the time periods may differ (from days to months to years to generations and so on). The impulse patterns from different impulses may be "superposed" or added together to a certain extent.

THREE CASES

1) American Revolution

The "mountain" view of the American Revolution, with a central decision point about October 1779 (at t=2 relative to t=0 as the Saratoga victory October 1777)) is presented as Figure 1 (=Graph 1AR) from a previous study. Here is added (Please see Graph 2AR.) a "foothill" impulse about t=-1.83 or December 1775, i.e. 1.83 years before October 1777 (Two years before would be October 1775.). At this decision point, mostly favorable to the British side, the Colonists failed to capture Quebec, Canada Dec. 31, 1775, the British King George III decided to consider Massachusetts as normative (rebels) versus colonists who wanted reconciliation, and Thomas Paine wrote "Common Sense," confirming the separation of the Colonies.

The plus-and-minus two-period points of Dec. 1773 and Dec. 1777, supposedly favorable to the Colonials, saw unopposed Colonial action of the Boston tea party, Dec. 1773 and the taking prisoner of the Burgoyne's army after October 1777, starting to bring France in as ally of the Americans.

The British reaction to the Boston tea party of the Intolerable Acts of Spring 1774 saw a one-year advance of Colonial opposition July 1774 to July 1775, including the Continental Congress leading to the Battle of Bunker Hill June 17, 1775, where British losses, together with further fortifications, forced the British to abandon Boston.

The significant thing from the point of view of this study is that two limited peaks (one from each of the two impulses) coalesce (Please see Graph 3AR.) to give a platform (although not that high) favorable to the American of about 9 months from April 1776 to Jan. 1777 (which looks something like the Rock of Gibraltar on the graph). Such a long period favorable to one side does not occur anywhere in the 1-impulse model, so that it represents a new emergent phenomenon of the 2-impulse model.

Of course this time interval had very difficult moments for the Americans, which requires the third impulse at the one-month (or 13.5 27-day periods per year) level, centered about 1.05 time units before October 1777, or say Sept 15, 1776, when the British occupied New York City. However this interval also saw the Declaration of Independence, July 4, 1776 and 2 out of 3 of George Washington's battle victories, Trenton Dec 26, 1776 and Princeton Jan. 3, 1777.

Some of the events of this interval April 1776 to Jan. 1777, based on the sum of all three impulses, are plotted on Graph 4AR.

There is a symmetric "Rock of Gibraltar" interval after Oct 1778 favorable to the Americans on the other side of the Saratoga (t=0) point.

2) Theodore Roosevelt Presidency

First an interesting point is the extent to which the life of Theodore Roosevelt follows the pattern of the "Conflict in History" theory, as illustrated in Graph 1TR. This result is compatible with his goal of the strenuous life "in the arena." Theodore Roosevelt had a sickly childhood after birth in 1858. He gradually built himself up physically and had some literary success after graduation from Harvard, but had a difficult start at a political career, which ended temporarily when both his wife and Mother died on nearly the same day in 1884. There followed a 25-year advance, starting with work as a rancher in the West and return to politics leading to nearly two terms as U.S. President 1901-1909. An award of the Nobel Peace Prize in 1910 was followed by defeat as Presidential candidate in 1912 and decline to his death in 1919 at age 60 short years after contracting jungle fever in the Amazon

In this case the "mountain" conflict seemed to come first, with his charge up San Juan hill at the head of the Rough Riders in the 1898 Spanish-American war considered as the high point of his life (Please see Graph 2TR.). The decision point of this first conflict seems to be centered on March 1900, perhaps representing a decision of making his mark in the new century. A second decision point seems to take place around March 1903, when he bests Mark Hanna for the Republican Presidential nomination for 2004 (Please see Graph 3TR and 4TR.). This impulse seems to be the point when John Singer Sargent painted his famous 1903 Presidential portrait of Theodore Roosevelt on a staircase. The two impulses registered much cancellation as his progressive goals conflicted with his promise to maintain a conservative stance to follow up on the assassinated (1901) McKinley's policy. Please see Graph 3TR for both one-year scale impulses (and including the one-month scale impulse at t=-1.5) plotted individually, without superposition. Thus in becoming President and during his "accidental" first term, he seems almost to move sideways, over many setbacks. Please see Graph 4TR for the superposed version, including the one-month impulse.

The third impulse at the one-month or 27-day period can be taken as centered at t=-1.5 or one and one-half years before March 1900 or Sept. 1898 and with a minus one (-1) coefficient. Please see graph 5TR for an expanded view of the one-month interval, including Theodore Roosevelt's "crowded hour" at San Juan hill. A finer scale at one-hour could be studied. Observe in Graphs 3TR, 4TR, and 5TR the three impulses are plotted with a minus sign; this minus coefficient is generally associated with aggression. Compare the three impulses in the American Revolution, with the Patriots on the defensive, are all plotted with a plus sign. Three peaks with the minus one multiplier occur within three time periods, or five peaks within five time periods, so that a "platform" is not as necessary to get significant results, as with the plus one multiplier for the impulse with two large peaks within four time periods. A quick review:

July 1, 1898 2 ½ periods before Sept 6	successful charge up San Juan hill
July 14, 1898 2 periods beforeSept 6	problems getting troops medicine/food
July 28, 1898 1 ½ period before Sept 6	circular letter to get troops out of Cuba (before Aug.3, troops leave Aug 7)
July 28 to Aug 24 one-period decline	men very sick on transports/Montauk

Aug 24, 1898—1/2 period	Problem with Chapman Indep. party
Sept 6 0 = impulse decision point	Sept 4 shakes hands with McKinley at Review; Republican Contender for Gov
Sept 20, 1898 ½ period	Problem with NY Residency
Sept 20-Oct 17 one-period advance	organizes Special Train as candidate
Oct. 30, 1898 2 periods after Sept 6	health problems/exhaustion on train
Nov 13, 1898 2 ½ periods after Sept 6	declared winner/NY Gov by Nov 8.

3) Viet Nam War

The Viet Nam War can be studied in terms of one "mountain" graph (Please see Graph 1VNW, with North Viet Nam success plotted above the horizontal axis and U.S. success plotted below the horizontal axis, to show the similarity with the American Revolutionary War.), based on a decision point of Feb. 1971, close to the repeal of the Tonkin Gulf Resolution (Jan. 13, 1971) and congressional limitation on U.S. troops in the Viet Nam area. The goal seems to have been to obtain "Communist containment" with "no more Dien Bien Phu," i.e. no American army capture, similar to the French defeat at Dien Bien Phu in 1954. At the two-period points Feb. 1969 and Feb 1973 favoring North Viet Nam there are the nearby-in-time deaths of Pres. Dwight Eisenhower March 28, 1969 and Pres. Harry Truman Dec. 27, 1972. Also the Paris Treaty favoring North Viet Nam was signed Feb. 1973 over the protests of South Vietnamese leaders.

Following the U.S. moon landing July 20, 1969 by Neil Armstrong at a point favorable to the U.S. and the death of North Viet Nam leader Ho Chi Minh—who looked like a combination of Fu Manchu and Emperor Ming of the Flash Gordon TV series —Sept.2, 1969 there was a one-year advance of North Viet Nam interests, due to continued U.S. bombing halts and increasing protests against the war in the U.S., including those in Washington D.C. Nov. 15, 1969 and those provoking the Kent State shootings May 4, 1970.

The one-year advance of U.S. interests Aug. '71 to Aug. '72 saw the Nixon administration resume bombing and somewhat outmaneuver North Viet Nam diplomatically by Nixon visits to China Feb. 72' and Moscow May '72, while cutting casualties by removing U.S. ground troops, leading to his re-election Nov. 1972.

There are close similarities with the U.S. Revolutionary War. At the supposed minus-3-period point favorable to North Viet Nam was the "Tet" offensive Jan.-Feb. 1968, parallel to the Washington campaign around New York Sept. 1776 (Please see Graph 1AR.). The "Tet" offensive contributed to Pres. Johnson's decision not to run for re-election March 31, 1968, favorable to North Viet Nam. At the supposed plus-4-period point favorable to North Viet Nam was the fall of Saigon April 29, 1975 and departure of U.S. troops parallel to the departure of British troops from New York City Nov. 25, 1783 (Please see Graph 1AR.). Another parallel is the rewarding/promotion of failed generals.

Curiously according to this model the development of the "Conflict in History" theory by the author into summer 1971 took place at one of the "future points."

Among events that do not seem to fit the theory is the Forrestal aircraft carrier accident July 29, 1967 at a point supposedly favorable to the U.S.

To extend the "one-mountain" version to a "foothills-versus-mountains" model requires a determination of a second earlier impulse that led to the large-scale U.S. war in Viet Nam. An approximate model is developed here in Graphs 2VNW, 3VNW and 4VNW. The decisive point is taken as the 1964 election between Johnson and Goldwater Nov. 1964, i.e. the Viet Nam War is considered a creature of the 1964 election campaign.(cf. Johnson's NSAM 273 of Nov. 26, 1963 vs. Kennedy's NSAM 263 of Oct. 11, 1963). To counter the "soft on communism" attack of Goldwater, Johnson had engineered the Gulf of Tonkin Resolution August 7, 1964 before the election. Johnson apparently planned to make peace with Ho Chi Minh after the election, but couldn't. leading to an escalation decision over the next year to July 1965. There seems to be controversy among historians over exactly when the decision was made. There is also the possibility that the fundamental decision was that of North Viet Nam to invade South Viet Nam.

Anyway putting decision points at plus-and-minus 3.2 years around Jan.-Feb. 1968 (i.e. Nov.-Dec. 1964 and March-April 1971, leads to a "platform" model (Please see Graph 2VNW.) similar to that for the American Revolution (Please see Graph 2AR.), only with one "platform "at the Tet offensive instead of two as for the American Revolution case. Putting the second impulse at plus 3 periods (versus 3.2 periods) consistent with the "one mountain" model leads to a more peaked "platform." The platform consists of a combination of the plus 3-period peak of the first impulse with the minus -3 period peak of the second impulse.

According to this model 2VNW, North Viet Nam had a one-year period of advance from June 1963 to June 1964. This period saw Kennedy falter/stumble in his plans for involvement in Viet Nam, leading to the Diem assassinations Nov. 2, 1963, and the Kennedy himself assassination Nov. 22, 1963 and a series of unstable governments in South Viet Nam. The resulting confusion brought North Viet Nam support from both the Soviet Union and China and confidence to reject the Johnson attempt at peace.

Of course the supposed one-year U.S. advance from June '65 to June '66 resulted from the U,S. escalation.

As with the American Revolution study, the 2-impulse model can be extended by addition of an impulse of the 27-day-period,type, here with the 27-day or one-month impulse centered at the Tet offensive, similar to the Washington New York campaign. These possibilities are indicated in Graph 3VNW and Graph 4VNW.

CONCLUSION

Much work needs to be done in the Foothills-Versus-Mountains study. The combined "Foothill-Mountains" models can appear comepletely unrelated (cf. Graphs 2AR, 4TR, and 2VNW), even though the "mountain" components are basically the same. Getting the "foothills" part of the model correct does not necessarily lead to getting the "mountain" impulse correct. Physically, some important cities are on a platform or plateau between foothills and mountains. Similarly how the foothills and mountain contributions may interact by causing a plateau favorable to one side seems to select the winner in some important cases.

EARLY YEARS (TO PH.D.)

July 15, 2011

FAMILY

I (Dennis Glenn Collins) was born June 26, 1944 in Gary, Indiana at the Methodist Hospital, slightly less than one year after my sister Martha Lynn Collins (July 7, 1943). Both my Father Glenn Collins (August 15, 1897-Dec. 1, 1981) and Mother Irene M. Collins (June 2, 1909-July 6, 2004) came from large families and were teachers, so that it seemed people including relatives and former students were always visiting our two-story wood-frame house, the second house past the railroad tracks on IN 51, now 7108 Grand Blvd., going toward Hobart, Indiana about 3 miles to the north. As I recall My Father had bought the house and half acre for $600 after World War I from Johann Wojahnn. My sister is still living there. During my early years, when the wind was from the south, soot would stream across the railroad tracks from the coal-powered steam locomotives, covering any clothes left out on the clothesline to dry.

My Father came from a family of five brothers and sisters: George Henry Collins (born Feb. 18, 1889), James Daniel (April 10, 1891-1972), Rose Elizabeth (August 8, 1895), Glenn (August 15, 1897), and Hazel Lenora (Sept. 13, 1900). A son John (1892) died at age 5 from diphtheria, but the family took in a small niece, Kate Stillwell, at age 3 to raise. My Father was the greatest person I ever knew and could do almost everything, including wiggling his ears and juggling. He had trained on a minesweeper after joining the Navy in World War I and attended naval instruction classes at Harvard University.

My Mother was the youngest of a family of seven brothers and sisters: Ralph Edward Richman (Feb. 16, 1891-Oct. 16, 1972), Carl Louis (June 17, 1893-June, 1972), Luther Anton (Jan. 18, 1896), Mary Margaret (March 27, 1899-1976), Paul George (May 5, 1903-1963), Florence Janet (June 7, 1907-Oct. 10, 1995), and Irene Martha (June 2, 1909-July 6, 2004). My Mother attended a German Lutheran parochial school and had very high standards. She mentioned that once she was selected to lead a morning school hymn "Von Himmel Hoch (From Heaven on High)" and she started it too high, so that the director whispered, "That's a little too Himmel Hoch, My Dear," and she had to start over.

I have no memories of grandparents; My Father's Mother, Mary Sophia Herrick Collins, born 1867, passed away Jan. 8, 1924 and His Civil War orphan-train Father Daniel James Collins in May 7, 1929. Mary Sophia's Father Henry Herrick (passed way May 22, 1912) was born in 1835 in Willoughton, Lincolnshire, England, fought in Company H of the 46th Indiana in the U.S. Civil War, was invalided out after about two years, and married Elizabeth Gemberling (passed away October 14, 1895) in 1863, then had six children William, Mary Sophia, Eva, Benjamin, John, and Lucy Ann. My Mother's Father Charles Joseph Richman was born Nov. 10, 1869 and passed away April 1, 1936 and Her Mother Margaret Katherine Waltz Richman lived from Nov. 26, 1867 to Jan. 15, 1945. Relatives came from a village Lauterbrunnen in Switzerland.

While I was growing up, My Father taught grade school and was Head Teacher (effectively Principal) at Ainsworth Grade School, walking distance on IN 51across the tracks south away from Hobart, IN, and My Mother taught Latin and other standard subjects such as math at Merrillville High School about 5 miles west on 330 (old Lincoln Highway). My Father had a daughter Vallie from an earlier March 31, 1929 marriage to Helen Krueger, who passed away August 18, 1941 from heart problems. In 1948 Vallie married Serbian Dan Bodlovich, and they had five children, Mark, Danny, Helen, Sue, and Johnny. Thus we had much interaction with my half-sister Vallie's family. Her husband Dan worked for the EJ&E Railroad, and eventually became a buyer for the company in Chicago.

NEIGHBORHOOD

The first house after the tracks housed Tom Bowman and his wife Lydia. Tom was invalided out after being blinded in World War I, and was regarded as somewhat lazy by My Father. Lydia was descended from a pioneer family in the area and a repository of practical knowledge and all the latest gossip.

The third house (after ours, which included a lawn owned by my family between 2nd and 3rd houses) from the tracks north toward Hobart was occupied by several families over the years, including the MacAllisters, the Ed Snearleys, and the Jerry Sims before burning down while I was away. The lawn served for games of tag, croquet, and simplified baseball.

The fourth house7048 Grand Blvd. was a kind of brick mansion built by John Hecimovich, a bricklayer in the Gary mills, for his wife Marie and seven kids, Kathleen, Margie, Jack, Mike, Patty, Maureen and Mary Joyce, who were somewhat younger than me and my sister.

Our family often attended movies at the Art Theatre on Main Street in Hobart on the weekend when not travelling; mostly I looked forward to the cartoons, such as Bugs Bunny. I recall being terrified by a tiger in apparently a movie "The Man-Eater of Kumaon (1948) which featured a character Dr. John Collins being attacked by the tiger. I recall My Mother later sitting next to me on my bed and explaining that logically a tiger couldn't fit under the bed and attack me, which I grudgingly accepted. After one of the movies about age four I was leaving and pointed out some candy to the vendor behind the glass case; My Mother showed up and said, "Whoa, boy, you have to have the money to pay before you can get stuff." I don't remember getting the candy, although I might have. Our family often ate at the Abbott's Restaurant on Main Street in Hobart once a weekend, and I still remember looking forward to the apple pie a la mode.

GRADE SCHOOL

Although I never attended kindergarten, My Father showed me how to draw a simplified cowboy with an infinity sign with an intersection sign loop on top of it as the cowboy's hat when I was four or five years old and took me to school a few occasions to display this skill to other teachers. My grade school teachers were Mrs. Ziesneus, Mrs. Thompson, Mrs. Harper, Mrs. Walters, and Mrs. Bernard from first to fifth grade at the Ainsworth Grade School (which said W.J.Hahn above the door I recall), after which I moved to the Merrillville School system, where I had Mrs. Sanborn, a Civil Air Patrol flying enthusiast as sixth-grade teacher.

By chance or design, I never had My Father as teacher, although my sister did. My Mother taught me how to do divisions during the fourth grade, so that I could calculate a running average of my batting success during recess baseball games. Defense was not the forte of these games, so that I could keep up a .500 to .700 batting average by placing my hits and not trying to hit the ball as far as I could, which most kids did. There was no such thing as walks in these games as I recall, as the batter was expected to swing at about everything.

SUMMER TRAVEL

My Dad was very interested in geography, so that our family went on several camping trips during the summers. One trip was through the northern states to Glacier National Park, stopping in Missoula, Montana where my Uncle Luther was Dean of the School of Fine Arts at Montana State University, but nonetheless gave us a good tour of Glacier and even drove us to Idaho. Uncle Luther visited us in Hobart, Indiana often on trips around the country, bringing chocolate bars and silver dollars to us kids. Another trip was to Tucson, Arizona where my Aunt Florence resided with her chemist husband, Clem Olsen, who worked at University of Arizona and had helped develop DDT during World War II. Later Florence and Clem moved to Mesa, Arizona. Uncle Clem gave me a very good lecture on probability theory.

Around Tucson, after driving us to some desert mission and showing us the exhibits of rattle snakes, gila monsters and tarantulas, he took us out among the cacti and pretended to be lost, to see if we kids could figure out how to get back, which I found un-nerving. Another trip went to New York City, where my Uncle Ralph's son Charlie worked, and his wife Florence took us to Chinatown and up the Empire State Building. During some of the camping trips my Mom and Dad slept in the back of a station wagon and my sister and I slept alongside the station wagon on cots. I remember being tormented by mosquitoes and getting pretty cold in the mountains. Sometimes we had a tent. While I was in junior high we went on a trip to the Colorado Rockies summer 1957, which supplied term-paper material.

SCOUTS

My Father was institutional representative for the Cub Scouts and later Boy Scouts, which I participated in. One of my goals was to create an Indian head-dress, but I could never find more than a couple bedraggled feathers. The paper drives to collect old newspapers around the neighborhood and camping trips to local parks were all right but I barely advanced through the ranks. There was a bus trip to see Ernie Banks and the Chicago Cubs play at Wrigley Field in which the bus broke down on Stony Island Ave. on the way back

COMMUNITY AND CHURCH

My best friend was Cal T. Shearer, son of a local merchant, who had a coal/oil supply store across the tracks south on IN51 toward the Ainsworth School. Cal was a year ahead of me in school and sometimes played organ from the balcony in the Hobart Trinity Lutheran Church, which our family belonged to. It was really awesome to sit in the balcony and listen to the J.S. Bach music, but the church building on Main Street was later torn down and another building with no balcony put up more outside town. With other kids Cal and I went on many sledding expeditions and once just the two of us were crossing Deep River between our house and Hobart when the ice started cracking, and we were barely able to jump to safety.

There were many community activities organized in Ainsworth then, including card games and square dances. I remember a trip to hear WLS Captain Stubby and the Buccaneers in Chicago.

Junior High

In seventh and eighth grades, as well as at Merrillville High School from ninth to twelfth grades, students moved from classroom teacher to different classroom teacher according to subject. I was lucky to have Mr. Horner, a family friend for algebra teacher in junior high, and I liked to work through word problems looking for the correct "Let x equal . . ."

High School

This practice stood me in good stead in High School and I ended up Valedictorian with a 4.00 A average. My Mother served as my teacher for four years of Latin. Again I was fortunate to have good math teachers in High School, including Mr. Hutchinson in geometry, Mr. Vermillion second-year and Mr. Rainford third-and fourth-year. Mr. Rainford let me work sometimes ahead of the class in the algebra texts BOOK ONE and BOOK TWO by A.M.Welchons, W.R. Krickenberger. and Helen R. Pearson. Working on my own was related to the Indiana University High School competitions, held in Bloomington, Indiana, which I entered each year, first and second years in Latin I (freshman year) and Latin II (sophomore year), and then third and fourth (junior and senior) years in mathematics. Every year I won at least a bronze medal in these competitions, and a couple silver medals I believe. I switched from Latin to math competitions after two years because the Latin words and constructions were somewhat arbitrary to remember versus the math problems. It was necessary to win at a regional level before getting to the finals at Bloomington. My Mother never complained about my switching to math although my winning in upper-level Latin would have contributed to her prestige as a Latin teacher. My winning the state medals cut down on the ribbing I got as somewhat of a "grind." Also during High School I took up piano lessons again with Mrs. Jackson of Hobart after earlier lessons with Mrs. Semokaitis I believe and played Chopin's Etude Opus 10, Number 3 in a recital. Athletically and socially I didn't do much, although I was really good at ping pong during lunch hour. I went out for wrestling, which was a big help to my upper-body strength; however my arms were getting so many muscles I had problems taking notes. Also I had a problem remembering my name for a few minutes after wrestling a classmate Klaus Ackermann, so that I decided I better stick to what I was good at, namely academics. I never attempted to a part in a class play, or went to a prom,

After High School in the summer of 1962, My parents got me a job at Jansen's fruit/produce market at Route 6 and IN 51 north of Hobart. Some rich people came there from summer homes around the Indiana dunes and I remember waiting on Dale Messick, who authored the cartoon "Brenda Star" and had a blue thunderbird convertible. The work wasn't too hard except unloading 100-pound sacks of potatoes and later cleaning out rotten potatoes. The owner, Cecil Jansen a Swede was pretty careful to see I didn't get hurt. However my nose had problems recovering from unloading crates of onions from above in the freezer, wherein the onion juice would drip down in your face and also get on your hands and everywhere when you cut the green ends off. I recall the market workers debating what had caused the death of Marilyn Monroe, August 5, 1962. At least one older member of the Jansen family had terribly arthritis-deformed fingers from a lifetime of picking moist beans and so on in the fields. Cecil's son Wally told exciting tales of driving the produce trucks back from Michigan.

COLLEGE

September 1962, I started on full scholarships in the Directed Studies Program as Freshman at Valparaiso University fifteen miles to the east of IN 51on U.S. 30, staying in Wehrenberg Hall my first year. Directed Studies students got to interact with the awesome President O,P, Kretzmann a couple occasions. The Directed Studies Program was a pre-cursor to Christ College being set up as a Honors College a few years later. Princeton University had admitted me as student, but without funding. My Brother-in-law's boss in the EJ&E Railroad was a Princeton grad and offered to help me attend, perhaps in return for working with the company later, but I decided not to follow up this possibility. Most likely I would have been chewed up and spit out by the Princeton crowd, as the high-school class Salutatorian (2nd ranking student) attended MIT in 1962, but couldn't stay there, which somewhat foreshadowed my one-year grad-school stay at Yale University 1966-1967.

`Going to Valparaiso University overall proved a good decision. Again I was lucky with my math professors, having the Math Dept. chairman, Dr. Kermit Carlson, as first-semester calculus professor. I had trouble following the official textbook, but was able to do well by reading the text by Thomas in the library, a version of which still seems to be in use (2011) at Purdue North Central, the local branch of Purdue University in northern Indiana. Malcolm Reynolds was also very competent in calculus math, as well as John Lennes, the son of a more-well-known mathematician, in linear algebra. Going through linear algebra proofs from day to day was pretty easy compared to the proofs that I had in real analysis with Kermit's brother Lee Carlson. Classes with Dr. Marvin Mundt in applied math proved to be my favorite, although I also liked the statistics classes of Dr. Foster.

Physics classes got off to a good start my freshman year but I had problems starting the second year with finding a good text for mechanics. The books in the library didn't seem to agree on the formulas. The quantum mechanics class with Dr. Armin Manning was fascinating but I had difficulty finishing the labs and ended up with a D in the class. Drs. Manning and Bretscher worked on nuclear submarine projects I believe. About five years later in graduate school I finally went back and completed several of the labs and turned them in, perhaps a record for turning stuff in late. I never heard anything back about them.

VALPO UNIVERSITY EUROPEAN SUMMER TOUR 1963

As a reaction to the 1957 Sputnik, Valparaiso University had started Russian classes, which I took during my first two years at Valpo. Getting across the frozen campus "tundra" for these early-morning Russian classes with the professor Robley Wilson furnished a good background for Russian language study. The polished lectures of the Western Civilization classes by Dr. Willis Boyd in the Directed Studies Program proved a good companion course, so that, thanks to my parents' money, I eventually signed up for the "Valparaiso Student Tour of Europe and the USSR," from June 19 to August 14, 1963 for six credits. The tour went mostly well, although I somehow missed the fact that tour members were supposed to be keeping journals. However I collected as many postcards as feasible. Although missing Spain, the tour made a nice clockwise circle of Europe from Britain to Scandanavia down through the Soviet Union and back west through Turkey, Greece, Italy, Austria, Switzerland, Germany and France.

At the Kasbah in Istambul I bought a Turkish officer's sword for $10, which I shipped back to Indiana from Rome, unfortunately the sword arriving somewhat bent. Perhaps as an early example of sports taunting, most everyone cheered when the tour got out of the Soviet Union to the airport outside Sofia, Bulgaria. As I recall the plane was stopped on the tarmac for several hours surrounded by soldiers with machine guns before proceeding.

SUMMER STUDY 1965 AT ILLINOIS INSTITUTE OF TECHNOLOGY

The summer after my junior year in college, I studied a few weeks in I believe an NSF program with Dr. H. L. Pearson, a highly-competent math professor from Canada, at Illinois Institute of Technology in Chicago, going through the book *Special Functions* by Earl Rainville as well as a small book by Harry Hochshild and a book on differential equations, which I really liked. There were pick-up softball games in the afternoon outside the dorm where I stayed, which I also enjoyed.

SUMMER WORK AT U.S. STEEL CORP., GARY, INDIANA 1966

Before graduating I was admitted to Yale Graduate School in the Mathematics. Department, starting Sept. 1966. However during the summer of 1966 I worked on an Industrial Engineering project measuring output at U.S. Steel Gary Works in the chipping yard. The chipping yard was charged with cutting seams or cracks out of "rounds," basically thick axles of steel about 30 feet long, mostly destined to be car axles after further processing. Sometimes the cracks could be seen by eye but also there was a "Magnaflux" machine which marked the cracks with fluorescent dye before the chipping process, done with pneumatic chisels. The pollution was terrific. Typically a dark cloud of pollution hung over the city of Gary, Indiana. Going into the plant, there was an even denser layer of pollution due to steel-making furnaces. Finally inside the enclosed "chipping yard," the smoke was even worse, filled with tiny metal particles from the chipping. Coughing into a handkerchief after work, it would come out black. The work was shift work, which meant the hours shifted from 8 a.m.-4 p.m. shift, to 4 p.m.-12 midnight shift, to 12 midnight—8 a.m. shift from week to week. There were lots of rats on the loose and the mostly huge black workers would hang their lunches overhead from strings. To get to work and back, my parents helped me buy for a couple hundred dollars a rusty 1959 Chevy, which I later called the "aircraft carrier" because of its tail fins. (The model appeared in a "Mad Max movie.") Its main problem was that the floorboards were rusted out, so that carbon-monoxide laden fumes would come up at the driver. During those years there were over 100,000 workers at the steel mills around Gary.

NSF FELLOWSHIP AT YALE GRADUATE SCHOOL 1966-1967

Summer work left me drained but I with my old Chevy I got on U.S. 30 which went from near Hobart, Indiana all the way East to within a few miles of New Haven, Connecticut., where Yale University is located. The trip was mostly uneventful but I remember going around in a loop of tens of miles in a National Forest in Pennsylvania, apparently because being dis-oriented by the fumes in the car. I arrived a couple of weeks before classes started and had to find a place to rent. I thought I had agreed on a place to rent one Friday but found out it had been rented to someone else when I checked on Monday, an un-nerving encounter with Eastern ways. I finally rented an apartment at 91 Lake Place upstairs from an old Swedish couple, although the rent took much of my money. There were other Yalie renters of Colin Bell and Charles Kelly downstairs, I recall.

Because there was still a week or so before classes, I decided to take a quick tour of Rhode Island and Massachusetts. The tour went mostly all right and I visited Brown University in Providence, Rhode Island, where I bought a copy of the highly-abstract Jerome Spanier's book on Algebraic Topology for $15, and went though the Harvard Campus in Boston. Although I hadn't really wasted any money, I found my self short of cash and wondered if I could make it back to New Haven without running out of gas. I stopped at a gas station and temporarily traded my copy of Spanier for $2 worth of gas, which the helpful owner assured me would be sufficient to get back to New Haven. I remember him musing that maybe he could pick up business a little by studying math. Later I sent him the $2 plus postage and he kindly returned the Spanier book by mail.

There was another problem. I had assumed the bank would cash my check from the Hobart bank; however the bank said not so fast; first open a bank account and wait until the check clears. Thus I would go to the bank each day and the clerk would joke that maybe they were sending the money by Pony Express. Meanwhile my small amount of food that I had bought ran out and I had to go a day or so without any food before the money came through.

Eventually the NSF money came in, the graduate-school cafeteria became available and classes started. However things got worse. During my whole month in New Haven I had not had to make any appointments, so that I had never noticed that the Eastern time zone was one hour ahead. As a consequence I arrived for my first class, incidentally with my NSF advisor, just as the class was ending. Also during the class I missed, the facts that there were help sessions available and homework due were announced, which facts never reached me. Thus I found myself struggling more and more in the Abstract Algebra class as my zero homework record advanced. At Valparaiso University the Math Dept. had used the algebra text by George Mostow from Yale, which text I liked, but he was on leave the year I stayed at Yale. Yale used the older algebra texts by Nathan Jacobsen, who it seemed was the Math Dept. Chairman at Yale, although in Spring Semester 1967, I also recall Shizuo Kakutani as being Math Dept. Chairman and sneezing in his office on the upper floor, which could be heard around the building.

It seemed my last change to catch up went out the window when I returned to Hobart for Thanksgiving, involving long and tiring rides and waits for the plane trips and no access to the Yale math library, the only source of the French Bourbaki volumes then in fashion at Yale. I hadn't connected very well with the other students since I was mostly accustomed to working on my own, although the conversation was often lively among the graduate students at the dorm cafeteria. My attempts to interact with the women graduate students came to nothing, although I did manage a couple dates. From current knowledge I would say Yale algebra was on the wrong track for my purposes and I wish I had had the chance to study the work of Bruno Buchberger on so-called Grobner bases, which revolutionized algebra in a more mechanical way.

Another problem was that my car was stolen, although later found abandoned a few miles away, no doubt after the thieves were overcome by the fumes. Nonetheless checking with the police was a distraction. Eventually I sold the car for $65 I recall, as the battery kept running down and it couldn't be driven to classes anyway. If I had known about car repairs, I could have plugged up or replaced the floorboards and possibly saved my Yale career. It was quite a ways walking, getting from my apartment to the Math Building. There was a guy living in a junk car behind my apartment well into the winter.

The second semester of the academic year I tried to escape the Yale math scene somewhat by taking more courses in other departments, including linear programming in the Industrial Engineering Dept. and solid state physics in an applied physics department. This was probably a mistake as far as passing the courses went, since I was spread even more thinly. Colin Bell, who happened to be in Industrial Engineering kept up my spirits some with his New Orleans jazz hobby, and as I recall Charles and I got a ride to visit Smith College one Saturday.

I managed to stick it out the rest of the academic year at Yale, although I'm lucky my parents came and got me at the end of the semester. Some evening lectures I attended on Einstein's later years by Cornelius Lanczos were the most interesting activity of this period. I think it was a year or so after I left that Yale started to admit women undergraduates. However somehow I filed to transfer my NSF Fellowship back to Illinois Institute of Technology for 1967-1968.

Graduate Study at Illinois Institute of Technology 1967-1970

Although I got off to a rocky start, moving back into the graduate dorm Sept. 1967 at Illinois Institute of Technology (IIT) in Chicago, it was nice to be working on more tractable math courses, such as Shu's course on non-linear differential equations. It's pretty hard to recover from bad experiences in math learning, and I was fortunate to be able to keep at it, finishing an M.S. in June 1970 with my thesis "Aggregate Logic" under H. Ian Whitlock. The IIT Math Dept. had downsized and Dr. Whitlock was my third M.S. advisor. I had first worked with Dr. Brender on topology on finding the covering spaces of two-dimensional compact manifolds, and after considerable effort, had worked out all the cases geometrically. Brender had said "That's good but you haven't done anything yet." Brender was very likeable and competent in algebraic topology but, as I heard, some of his research papers were rejected and he left IIT. My fellowship provided $2 per month more than my room-and-board expenses, which theoretically allowed one bus ride to downtown Chicago and back recreation per month. Once I had an unknown hole in my pocket and my return money fell out, but a kindly pharmacist lent me the money, 50 cents I recall, later repaid, for the bus ride back to IIT at 31st Street. Walking it would have been possible but pretty dangerous. Sometimes I took the South Shore electric rail line home to Gary, Indiana, where my parents would pick me up for the weekend at Hobart and My Father would pay me a little for doing chores. Relatives would hint, "When are you going to get a job." In 1968, I happened to be on the way to Hobart the day Martin Luther King was assassinated and got a taxi ride from Gary most of the way home rather than have my parents come to Gary.

Ph. D. at Illinois Institute of Technology 1970-1975

To continue math studies toward the Ph.D. at IIT, I had to start teaching as a ½ to 1/3 part-time teaching assistant. My first class as instructor was a beginning algebra class at 33rd and Federal, next to the Chicago L-tracks, which evening trains went roaring by right outside the window of the classroom occasionally. The class went well I thought, perhaps partly due to helpful suggestions by one of the better students, Mrs. Heffernan and her husband. Later I had as instructor several classes of Math 418 Partial Differential Equations with the text by Ruel Churchill, which text I had studied extensively, going back to the 1965 summer program at IIT.

Meanwhile I had continued my questionable practice of taking advanced courses in other departments, and had run into a couple texts that really interested me, such as in an Electrical Engineering course, *Communication Theory* by B.P. Lathi. Of course, such interest partly depends on the relevant professor knowing and explaining the material well. Eventually I took advanced courses in seven departments. Although some students stayed in school to avoid the draft during the Viet Nam War, this dodge was not my motivation. I was called up for the Army physical exam and went through the procedure, at length being classified 2-Y, retrainable in a national emergency, which classification I liked. As a consequence of my Yale experience, and to orient me to future direction, My Mother had set up some appointments with a psychologist B.J. Sperov, who happened to have graduated from IIT, and was very helpful encouraging me. Anyway he wrote a message saying he thought it would be better for me to continue my studies at IIT. On the bus to take the Army physical exam, I found that, unlike me, most of the fellow riders had medical notes trying to help them get into the Army, as there was a surplus of potential inductees from Lake County, Indiana, due to the Army being a way to get out of poverty. As I recall there was one Friday, perhaps between the Army physical and the classification, where I started home late on the South Shore to Gary, and panicked when I realized I didn't have sufficient money to get all the way to Gary and also call my parents, who might be less than enthusiastic to pick me up in Gary so late at night. Thus I got off partway from Chicago to Gary, to evaluate the situation, while I still had the money to get back to IIT. Somehow there were no trains between 2 a.m. and 5 a.m. and the station closed down, with me staying outside in a phone booth while it started to snow. Before 5 a.m. kids with duffel bags started arriving out of the darkness, who were on their way to Army basic training on the South Shore, and asked, "Are you going—" I said no, and later went back sheepishly to IIT.

At length I had the required course credits and had to pass the comprehensive exam and do the dissertation to finish the Ph.D. I passed the written comprehensive tests all right but had to take the oral comprehensive test twice to pass. One the committee members had said to me, "You may be ten Einsteins, but we can still fail you. We can go to the footnotes if we have to." Anyway the committee asked me to prove an obscure measure theorem involving spheres, which I couldn't do. However after studying the offending theorem, which is actually quite interesting, and a few other recommended topics, I passed the second try.

My second M.S. advisor had been in the computer science department, so that I returned to these questions for my Ph.D. dissertation on "Preference Matrices." My Ph.D. advisor Dr. William Darsow was really friendly and knowledgeable, and I liked his lectures on group representation theory, I believe. Dr. Darsow was not so much an expert on computers, so that he arranged to have the relatively well-known expert Prof. C.L. Liu of the Computer Science Department at U. of Illinois, Urbana-Champaign evaluate my work. Some part of the work involved solving a linear programming problem, submitting decks of cards to the IIT mainframe. Anyway, the dissertation was approved in early 1975. Actually a major stumbling block finishing the dissertation was getting it printed, since it involved a large number of mathematical symbols. In the end I bought a set of metal mathematical symbols and laboriously stamped them on the pages one-by-one. By the last page I recall my fingers were bloody from the sharp edges of the metal type.

Then in June 1975, I received the Ph.D. in Mathematics from IIT. My Uncle Curt attended and gave me $50, a large sum of money to me.

Earlier I said my trip to Yale in 1966 was mostly uneventful. I guess I repressed the fact that I had stopped to visit my Aunt Sis, as My Father's favorite sister Rose was called, with her husband Curt, who lived in Akron, Ohio close to U.S. 30. Curt was a successful businessman with a tire company and also was a Mason. As I was leaving, across the street to continue to Yale, My Aunt Sis slipped on the curb in front of her house, fell down and broke her hip. Thus there was a delay as she was taken to the hospital, and although she recovered somewhat, she was never able to get around again without a walker. Today she would probably be able to walk fine with an artificial hip after such as accident. Anyway it was great that my Uncle came to my graduation and even gave me money.

Over I years I have tried to publish my dissertation a couple occasions. In 2010, I tried to give a talk on the work at a logic conference in Chicago, but there was some complaint about the symbols in the abstract and it was rejected before I had a chance to reply. I figured the dissertation was 25, now over 35, years ahead of its time in 1975 but the technology is catching up and maybe already has caught up. My interaction with Computer Science at U. of Illinois, Urbana-Champaign I think contributed indirectly to the proving of the famous 4-color map theorem there a few years later. At least I had felt the Yale expert Oystein Ore was on the wrong track trying to prove the question was unsolvable. I believed that there was only one maximally-connected case which map could be colored and if there was any deviation from that case, the number of possible different colorings would go up.

CONFLICT IN HISTORY 1971

One of the outgrowths of my electrical engineering and quantum physics projects was my work on conflict in history, based on the idea that the restrictions on real-life frequencies affect the type of events that occur, as for example the restrictions on light frequencies affect the colors that show up. What do I mean by frequency restriction—that people are creatures of habit and do the same things over and over, day after day, month after month, year after year, generation after generation. (For some reason I haven't got week after week to work out, perhaps due to too many vacation days.) Somehow the theory led to one formula, which has seemed to work from ancient history up to the present. The results were exciting and I looked up a Chicago copyright attorney Frank Thienpont, who helped me file the copyright application.

As with the dissertation, attempts to publish further got nowhere. One history critic said he did not know any reputable historian who subscribed to a cyclic theory of history. I wished to ask him if he expected the sun to come up and have breakfast tomorrow, and the next day, and so on, but I let it go. However my history professor Dr. Willis Boyd did give me significant support over the years until passing away on Oct. 17, 1993. The conflict in history theory may be dedicated to his memory.

Architecture Case Study in Transformity Factorization

Dennis G. Collins

Dept. of Mathematical Sciences
Univ. of Puerto Rico, Mayaguez
Box 9018
Mayaguez, PR 00681-9018
d_collins_pr@hotmail.com

ABSTRACT

This paper studies the Giannantoni factorization of H.T. Odum's transformity into dissipative and generative components. Emergy maximization is analyzed as a constrained calculus problem which for maximization requires middle values of both dissipation and generation. For example a placement of bricks around a yard in a highly symmetric fashion may have high symmetry but if they are not connected, will not lead to a desirable architectural structure. Similarly connecting the bricks into haphazard walls may have high dissipation but without some symmetry of construction into regular structures such as rooms, will be considered a waste of materials. Some other questions such as evolution of biological and animal structure are discussed.

INTRODUCTION

A dissipative component of architecture was developed in the author's paper "'Tropical' Emergy and (Dis-) Order" at the 4th Biennial Emergy Research Conference, and is related to the number of surfaces used up in architectural construction, for example making walls out of bricks. A generative component was developed in the author's paper "An Algorithm to Measure Symmetry and Positional Emergy of n Points," presented at the 2007 annual meeting of the American Mathematical Society, New Orleans, LA and included the the ISSS 2007 Bulletin; the generative component is related to the number of equal distances created between different parts of a structure. There is some evidence of ordinality; for example higher-dimensional structures can have orders of magnitude more symmetry.

There is a certain paradox in the search for maximum empower, in that emergy is based on used-up energy, so that it seems a more wasteful production process (with more used up energy) would be favored. A way out of this paradox may be offered by Giannantoni's factorization of transformity into a dissipative part, which is based on used up energy, and a generative part, which is based on creative molding of energy into new types. In this way a process which merely uses up a large amount of energy without creating any new energy type would not necessarily be favored. Since Emergy = Transformity *Energy, taking the derivative with respect to time yields the equation: Empower = Transformity* Power, supposing Transformity is constant for a given process. Then a process has greater Empower if it has greater Transformity, for a given power use. © 2008 Dennis G. Collins

To respond to a reviewer and explain why Giannantoni factorization makes a contribution to emergy theory, the factorization T=Td*Tg avoids the difficult mathematicla reality that otherwise confronts the theory: Consider the formula Emergy= Transformity*Available Energy. "Transformity" enters here in direct relation rather than inverse relation, i.e. as Transformity goes up, Emergy goes up. An efficient process apparently uses a smaller amount of Available Energy. If the Transformity also tends to a thermodynamic minimum, there is no way the the product can become a maximum, for maximum Emergy production, assuming this result to be the main rule sought after. Somethiing has to be increasing to get a maximum. The "something" can be the Tg factor, so that dissipation Td can be decreasing to its thermodynamic minimum and still the product T = Td*Tg can be increasing, so that the overall product Transformity*Available Energy can become larger.

From his work, it seems that Giannantoni considers both Td and Tg to be included in the current values of transformity. Thus the current tables could be extended by further study to get the approximate values of component factirs Td and Tg and these could be included in future tables.

How to calculate Tg is a topic for further study. As another reviewer points out "A and B" is not symmetric to "B and A", since in the first case "A" gets top billing and in the second case "B" gets top billing. From space earth looks pretty well spherically symmetric. However up close there are mountains and oceans and so on, which greatly affect the devopment of life. On the other hand life as known is dependent on the more or less symmetric distribution of oxygen and carbon dioxide in the atmosphere. Life would be much different if oxygen were distributed in clumps like land and water, or even moving clumps. Technological advances with higher transformity like jet planes are highly dependent on higher levels of symmetry in machining parts. To what extent higher transformity can be reduced to component levels of symmetry remains to be seen. Much technology depends on obtaining higher levels of purity (i.e. symmetry) than found in nature.

The simple non-dynamic or static models presented here do not take into account feedback, which can cause specific sub-systems to violate the maximum emergy production rule by contributing to more emergy production elsewhere.

There are a couple relevant optimization problems, letting T = Td*Tg, where transformity T equals dissipative transformity Td multiplied by generative transformity Tg:

1) max (T) = max (Td*Tg), such that Td <= A, Tg <= B,
2) max (T) = max(Td*Tg), such that Td+c*Tg = C. Here small c is a scale change between dissipative and generative transformity. and big C is a bound or limit.
3) max(T) = max(Td*Tg), such that Td and Tg are calculated dependent on specific placement of blocks.

1) The answer to the first problem is simply to take the largest possible value A of Td and the largest possible value B of Tg and multiply them, to get A*B.

2) By calculus, the answer to the second problem is Td = C/2 and Tg = C/(2c), so that the maximum product T comes out C^2/(4c).

3) The third problem remains challenging. In the case discussed below with ten 4x2 blocks, the maximization could be taken over the space $(R^3 \times U(1))^{10}$, where R^3 is three-dimensional Euclidean space and $U(1)$ is the one-dimensional circle of orientations from 0 to 2π, with the constraint that two blocks cannot occupy the same space. Although this setup is the official way to state the problem, it seems intractable at present.

There is some work on obtaining "minimizers" in the roto-translation space which the reader can consider (Citti and Sarti 2006).

However the first problem can provide a sound guide to solving the problem in the case of a local maximum.

The first problem is relevant if there is no "trade off" between Td and Tg, so that they can both be maximized at the same time.

The second problem is relevant if there is a "trade off" between Td and Tg, so that maximizing one of the factors decreases the ability to maximize the second factor.

In this paper the decision is made to measure dissipative transformity by the ratio of original surfaces of blocks divided by remaining block surfaces. If few surfaces are left, the denominator is low and the ratio (dissipative transformity) is high. Generative transformity is measured by the number of pairs of equal distances between blocks of the resulting structure.

This study is very limited in that it does not study the "Use Case," of architecture, or what the structure is designed for. These questions involve the placement of doors and windows and repeated pathways, which involve patterns, or cycles, in the time or fourth dimension. It is hoped later studies can extend the calculations to these essential questions. It would seem the most symmetry could be obtained in the 4-dimensional case by simply letting things stay the same from one moment to the next; however as H.T.Odum stressed and (Pico, 2002) points out, things are always decaying on their own, so that it takes feedback (change) to keep things to appear to stay the same, whereas they thus are not actually staying the same.

For the simple cases analyzed here, it seemed that problem 1) was sufficient to handle the architectural structures involved, i.e. that it was not necessary to worry about "trade offs," or, stated otherwise that it was possible, locally, to maximize both types of transformity Td and Tg at the same time.

MODEL CALCULATIONS

It was decided to limit the study to ten uniform 4x2 LEGO blocks. Each such block has an initial dissipative transformity, based on its being 8 cells (each cell with 6 sides) combined together, of 6x8=48 incoming sides dividied by 28 resulting sides (8 top, 8 bottom, 12 lateral), or Td = 48/28= 1.714.

The generative symmetry is based on putting one point at the center of each 2x2 block, or two points for each 4x2 block. The two points in each 4x2 block define an orientation. If the blocks are scattered randomly in space, the only equal distances will be the distance between the two points of each block. (Here the definition of random is that there are no more pairs of equal distances. This situation is somewhat hard to obtain; for example if any two of the blocks are parallel, it will create at least another pair of equal distances.) The minimum number of pairs of equal distances will then be the combination of ten things taken 2 at a time, or C(10,2) = 10*9/2= 45, since any two blocks create a pair of equal distances, based on the common separation of their two internal points.

As a consequence, the product T = Td*Tg will always be at least 1.714*45 = 77.14.

Now what about maximizing T-various configurations create local maxima; it is claimed these are the structures that appear in architecture. Small changes in these structures decrease BOTH the dissipative and generative transformity. The following structures are considered:

1) straight line (1-dim)
2) square outline (3 blocks across each side, like foundation outline of a square room) (2-dim)
3) solid rectangle (4x5, like a roof or wall) (2-dim)
4) solid bench (5 adjacent blocks on bottom level and 5 adjacent blocks on top level) (3-dim)
5) pillar (5 levels of 2 adjacent blocks) (3 dim).

There is a complication with 3-dimensional structures since LEGO blocks are not cubical; the side length of a 2x2 block is 5/8 inches, but the height is only 3/8 inches. This difference causes 4) and 5) above (both in some sense 2x2x5) to come out with different values. The results of the transformity calculations are as follows:

Structure	Td	Tg	T = Product
1) Straight line	480/244=1.967	1140	2242.62
2) Square outline	480/240=2.000	1163	2326.00
3) Flat solid rectangle	480/196=2.448	1992	4878.36
4) Solid bench	480/136=3.529	1052	3712.94
5) Pillar	480/112=4.285	1595	6835.71

Interestingly, any slight change in the above structures typically decreases both the dissipative transformity and the generative transformity at the same time. The following results indicate the result of the given slight change in the above structure.

Altered Structure	Td	Tg	T= Product	% decrease
1) Break line in two	480/248=1.935	1050	2032.25	9.3
2) Move one side over one unit	480/252=1.904	851	1620.95	30.3
3) Move one block out	2x2 480/204=2.352	1902	4475.29	8.2

4) Move one block out	2x2 480/152=3.157 983	3104.21	16.3
5) Move one block out	2x2 480/128=3.750 1446	5422.50	20.6

These results depend significantly on exactly how the pattern is broken, since the web of equal distances varies according to how exactly the block is moved; however the general idea will stay the same. In particular the case 2 in which two breaks were made in the square to move the side over has considerably more loss of transformity than the other cases with only one 2x2 block moved.

Remark 1: If cubical stacking were allowed (versus 3 to 5 ratio), the pillar case 5) would come out even more than the rectangle case.

Remark 2: The symmetry calculations Tg are based on one of the author's algorithms, for which a patent is applied for.

DISCUSSION

The results seem to follow the general outline of architecture, that certain structures—such as line, square, roof, pillar, and so on—representing local maxima of transformity, recur. Even slight deviations from these structures—a hole in the roof, a crack in the pillar—cause significant discomfort, i.e. decrease of transformity, mostly due to the generative (here symmetry) factor Tg.

There is also pause for thought in that the final transformities of the local maxima may not differ that much (4878 for roof versus 6835 for pillar), although any pathway from one to another (even straight line to square) may require an almost complete breakdown of transformity toward the minimum of 77.14 as structures are decomposed and re-assembled.

Thus the pulsing of one ecosystem or culture to another of nearly equal or greater transformity may go through the valley of chaos. This fact raises another question of how maximum empower might actually be achieved.

CONCLUSION

The simple results of this paper did not require the "trade off" theory of calculus; however more complicated cases would seem to require such trade offs. For example a house cannot be built only with a roof; it also requires pillars to hold up the roof. Thus the theory requires the further development of "Use Case" via time-varying or four-dimensional structures, to obtain practical results. It is believed, although results from one-to two-dimensional cases in this study only increased maximal transformity by a factor of about two, that higher-dimensional cases (such as four) may increase transformity by orders of magnitude. Only further study can determine if this possibility occurs. Also the question arises if symmetry per se can measure ordinal increases in transformity.

In terms of animals, the existing set of fauna seems to correspond to the local maxima of the architecture case study, as somehow giving at least local maxima of empower, in the sense that any small change in the structure of the animal is likely to cause problems for the animal.

The chart on p.588 (from Michael Land in Dawkins) depicting various evolutionary developments of the eye seems mostly compatible with the architecture structures considered here.

ACKNOWLEDGMENTS

The author thanks the Dean of the School of Arts and Sciences, Univ. of Puerto Rico, Mayaguez Campus for some travel money ($690) to attend the 5th Emergy Research Conference at Univ. of Florida, Gainesville, as well as Albertina Lourenci for discussions on architecture at the 4th Emergy Research Conference, and Mr. Willy Farrell of IBM for presentations on Rational Software Analysis at Univ. of Puerto Rico, Mayaguez.

The author thanks the organizers of the 5th Emergy Conference for allowing him to present this material and the reviewers for comments. If the paper is included in the Proceedings, it will be due to the efforts of Dr. Sharlynn Sweeney and others. Mostly thanks for Dr. Jennifer Wilby, a preliminary version of this paper was also presented at the 2008 Annual Meeting of the ISSS (International Society of Systems Sciences) Madison, WI and included in their Conference Proceedings.

REFERENCES

Alexander, Christopher 2002. *The Nature of Order, Vol.I*, Center for Environmental Structure, Berkeley, CA.

Citti, G. and Sarti, A. 2006 A Contact Based Model of Perceptual Completion in the Roto-Translation Space, J. of Math Imaging, Vol. 24, 307-326.

Collins, Dennis 2007. Algorithm to Measure Symmetry and Positional Entropy of n Points, *General Systems Bulletin, Vol. XXXVI*, pp.15-21.

Collins, Dennis 2006. "Tropical" Emergy and (Dis-)Order, *General Systems Bulletin, Vol.XXXV*, pp.17-22.

Dawkins, Richard 2004, The Ancestor's Tale, Houghton-Mifflin Company, Boston, p.588.

Giannantoni, Corrado 2002. *The Maximum Em-Power Principle as the basis for thermodynamics of quality*. Center for Environmental Policy, Univ. of Florida, Gainesville, FL, SGEditorial, Via Lagrange, Padua, pp.93,99.

Odum, Howard T. 1983. *Systems Ecology*. John Wiley and Sons, New York,

Pico, Richard 2002. *Consciousness in Four Dimensions*, McGraw-Hill, New York.

MORAL CODES III: SPIN AND REGULARIZATION IN JUDGMENT

By Dennis G. Collins
Dept. of Mathematical Sciences, University of Puerto Rico, Mayaguez
Box 9018
Mayaguez, PR 00681-9018

March 6, 2009

ABSTRACT: This paper continues the study of moral codes as developed by the author's *Thermodynamic Modelling of Moral Codes (Moral Codes I*, 2001) and *Moral Codes II*, 2003, 2005 to include spin and regularization. Although the conclusion of *Moral Codes I* remains valid, "that the moral code enables society to operate at a higher temperature than it would otherwise be able to do by disabling high energy 'bad guys,' "it is seen that this very process reduces the population at the tails of the moral code distribution, and thus causes the moral codes density to operate at a lower temperature (say 1.4) than it otherwise would (say binomial 4.5). A comparison is made with Puerto Rico crime rate. Spin is brought in to deal with the problem of instability of moral judgments as explained in *Moral Codes II (p.100)*: "The consequence is that almost any violent act can be justified from some viewpoint, and moral judgments become unstable, so that one person's level 0 (saint) is another's level 9 (mass murder). Regularization is a way of removing instability. The models are from generalized statistical mechanics of Gibbs, E.T. Jaynes, and Ising.

INTRODUCTION

The objective of this paper is to extend the model of the author's previous paper *Thermodynamic Modelling of Moral Codes (Moral Codes I*, [1] 2001) beyond energy and entropy to include magnetization and two "polarizations," so as to be able to describe more social science phenomena. In particular, according to the method of E.T. Jaynes [3, p.623] the maximum entropy density is based on the form

$$P(xi) = exp[-c+beta*(-E(xi)+h*M(xi)+g*L(xi)+v*P(xi))]$$ where c is a normalization

Constant related to the so-called partition function Z = exp[c] and beta = 1/T with T equal to the (abstract) temperature. Here E stands for the quadratic (based on level) social cost in Moral Codes I or the Ising energy (based on spin) in this model, M stands for magnetization (evidence

or propaganda in the social setting), L stands for left-right polarization (in the political setting) and P stands for vertical polarization.

The derivation shows that the above form of solution maximizes the entropy of the density subject to the constraints on *E,M,L, and P,* that they have given average values.

A future project is to combine the two types of energy. Here (*Moral Codes III*) in the Ising model it is the opinion (plus or minus spin) of 6 individuals that is of interest, so that there is no interaction with the unknown internal level of the individuals. Thus the applications of *Moral Codes III* are to judgments, voting and so on, about moral or political cases. There is no discussion about whether or not as part of a jury, a war veteran, say, who had killed people in a war, would be more or less likely to convict a person charged with murder. The subject of *Moral Codes III* is probability densities of judgments (for—plus spin, or against—minus spin) rather than probability densities of population moral levels.

MORAL CODES/UPDATE

According to *Moral Codes I*, at higher temperatures, the levels of the moral code are filled out until as (abstract) temperature goes to infinity, the density becomes uniform (probability equals 10%) over the ten levels, which are the following:

0) "saint," making everyone better
1) "Good Samaritan," making most better
2) Productive citizen, service making some better, but with limited risk to self
3) Decent person of good will, honors parents, not greedy but protects own interests
4) Law-abiding but has self interest with possible thought crime such as coveting
5) Word crime: swearing, lying, as well as thought crime
6) Property crime, stealing, cheating
7) Violent crime, assault
8) Murder
9) Mass murder, treason, crime against humanity, making holes in the dikes

Conversely, assuming a symmetric discrete Gaussian density (of form proportional to $\exp[-a*k*(9-k)]$, centered at average level 4.5, as temperature goes to 0, almost all the population occupies levels 4) and 5).

To the extent that the moral code (say police) tries to keep the population out of the higher levels, then it must keep (abstract) temperature low. Thus the moral code can be considered as a, say metal cylinder around the middle levels (4 and 5) keeping the population out of the extreme levels (0 to 2 and 7 to 9). Thus even though the goal of the moral code is to allow society to operate at a higher temperature in the sense that more levels of non-harmful activity can be occupied, as far as the moral codes density goes, the objective is to keep the temperature down, so as to limit disruptive behavior.

One problem of *Moral Codes I*, p.5 ("An interesting mathematical problem is to find out if there are choices for the temperature T, mean or average level mu, and cost of each level, such that the results of this model match the results of the binomial distribution.") can be more-or-less solved by observing that both the discrete Gaussian and binomial can be approximated by the Gaussian or normal density and therefore by each other. In fact the binomial with mean 4.5

and n = 9 trials and p= ½ can be approximated by a Gaussian with mean np= 9*(1/2) = 4.5 and variance npq= 9*(1/2)*(1-1/2)= 9/4 or standard deviation sigma = 3/2. Then the discrete Gaussian of form exp(-a*k*(n-k)) matches the Gaussian exp(-(x-mu)^2/(2*sigma^2)) if a = 1/(2*sigma^2) = 1/(2*(3/2)^2) = 1/(9/2)= 2/9 = .22 approximately, or T= 1/(2/9) = 9/2 = 4.5.

This discrete Gaussian (with a = .22 and T= 4.5) can be taken as normative of moral behavior without a moral code, namely, there are nine moral choices, each with probability ½ of being correct, and someone who makes no mistakes ends up at level 0 (saint) and someone who makes all nine decision wrong ends up at level 9. As stated in

Moral Codes I, the U.S. can be taken to have a temperature approximately T= 1.4 (or a= .7). In this view the effect of the moral code is to lower the (abstract) temperature by a factor of about 3 from 4.5 to 1.4. This decrease has the effect, for example of reducing the murder rate (level 8) from 1749 per 100,000 to about 8 per 100,000. See Table I.

It is possible to check what abstract temperature would correspond to the highest murder rates of about 17 per 100,000, which correspond to Puerto Rico [4, p.14] in 1998 (or 20/100,000 in 2002), or 19.5 per 100,000 in Dearborn-Livonia, Michigan (near Detroit) in 2005 [4,p.4]. A temperature of T = 1.5625 or a= .64 would yield a murder rate (level 8) of 17.7 per 100,000. According to this comparison, the "hottest spots" with failure of the moral code restraint in the U.S. only increase the abstract temperature from 1.4 to about 1.56. Actually the paper by Godoy [4] argues that the violence (level 7) and property (level 6) crime rates in Puerto Rico may be less than the U.S. mainland.

JURY/VOTING MODEL/SPIN

Now comes the main topic of the current paper, which is a model of judgments or voting by population members about moral questions. In particular it is observed that, working with symmetric densities, the energy is left unchanged by a spin flip or interchanging level 0 with 9, 1 with 8, 2 with 7, 3 with 6, and 4 with 5. Thus for a given moral problem, there are two states, a plus spin with levels 0 to 9 left intact, and a minus spin with levels interchanged. The model works with 6 individuals, so that with a 6-man jury, to convict all spins would have to be up, say indicating evidence, and to acquit, all spins would have to be down; otherwise there would be a hung jury. If initially three of the 6 thought the defendant were guilty and three thought the defendant innocent, then to avoid a hung jury, the spins of three of the six would have to change.

With 6 individuals, there are 2^6 = 64 possible states for the system, as each person could vote up or down. As mentioned in the introduction, a supposed Jaynes model, [3,p.623], is set up with four functions Ising energy E, magnetization M, left-right polarization L, and north-south polarization P.

Politically, the 6 units can be considered as six sections of a country, which can either vote plus or minus. The plus side can be considered to win an election if the probability of states with four or more spins plus (out of 6) adds to more than 50%.

The values of spin plus or minus one of the six individuals can be arranged as in a #6 domino, for example   1 1 -1

1 -1 1.

The Ising energy E is difficult to explain, but each individual spin is multiplied by nearest-neighbor spin and added; then the sum is taken over each individual. Finally the negative is taken. Diagonal relationship is not considered nearest-neighbor. For the above case, the top three individuals contribute the following: (1+1) + (1-1-1) + (-1-1).

Then the bottom row contributes (1-1) + (-1-1-1) + (-1-1) to give a total 2-1-2+0-3-2= -6, or taking negative E = 6. The upper left individual contributes 1*1+ 1*1= 1+1 =2 since he has two nearest neighbors, which are both spin 1. The upper middle individual has a neighbor to the left contributing 1*1=1, a neighbor below contributing 1*(-1) =-1 and a neighbor to the right contributing 1*(-1) =-1.

The left-right polarization L is obtained by taking a "dot-product" of the individual spins with the pattern -1 0 1 . For example the L-value for the above case would be -1 +0-1

     -1 0 1   -1 +0+1

=-2.

The north-south polarization P is obtained by taking the dot product with the pattern 1 1 1. The above case would yield 1 +1-1 for a P-value of +2.

  -1-1-1   -1 +1-1

The magnetization M is the easiest to calculate as merely the number of plus 1's minus the number of minus ones, i.e. the net sum. The above example would give M = 4-2 = 2.

The above values of E, M, L and P are extensive quantities determined by the given pattern of plus's and minus's. The values of beta= 1/T, h, g, and v are intensive quantities imposed from outside. For the social science setting, the value of magnetic field strength h could be considered as the amount of evidence (say, plus—for conviction and minus—against) or magnetism of the lawyer or propaganda. The value of g could be considered the amount of effort to polarize opinion to the left or right. The value of v can be considered as the amount of effort to polarize opinion on a north-south or top-bottom basis. An increasing value of temperature T, as outlined above, has the effect of spreading the probability over all the different states. A low value of T generally concentrates the probability into only a few states. If any intensive values are set to zero, the effect is eliminated.

The process of obtaining a verdict by a jury could be considered as similar to simulated annealing, whereby at first there are many possible states but the temperature is gradually lowered until one state dominates.

COMPUTER PROGRAM

A *Mathematica* program implements the Gibbs/Jaynes/Ising model described above. Please see Appendix I. The program calculates the probability of each of the 64 states for given values of beta (1/T), h, g, and v. A diagram illustrates the state and the probability is put underneath. For visibility, black is taken as plus spin and white as negative spin, although this selection could be reversed. A couple worksheets handed out as homework for the author's Mate4071.040 (Intro. to Methods of Modern Science II) class are also included.

RESULTS

The program worked as expected. A few examples are listed:

Jury decisions:
1. No evidence or bias. Low temperature. At T = 1/20, with other intensive variables set equal to 0, essentially all probability is concentrated on all plus spin (state 1) or all minus spin (state 2), each with probability ½. There is no way to decide the case.

2. Evidence against the defendant. At low temperature (T= 1/20) with evidence against the defendant (h =1) and other variables equal to 0, essentially all probability is concentrated on conviction (state 1).

3. Lack of evidence against the defendant. At low temperature (T=1/20) with lack of evidence against the defendant (h=-1), essentially all probability is concentrated on acquittal (state 2).

4. Stubborn jurors. If left-right polarization is set to g=2, with T= 1/2, and h= 1, and v= 0, there would be a hung jury since all probability is not concentrated on either state 1 (convict) or state 2 (acquit).

Voting behavior:

5. Suppose the electorate is polarized with g=5, so that the two sections to the left tend to vote minus and the two sections to the right tend to vote plus. Suppose debates and advertising favoring right (plus), so that h= 3. At temperature T= 1/20, the right would win either with all votes (state 1 has 50%) or with 4 votes (state 16 with 4 sections to right voting plus also has 50%).

6. At high temperature (say T = 5), there is more or less random voting. Most states have about the same probability.

Many other possibilities can be investigated. A typical case is listed as the output of Appendix I.

REGULARIZATION

Mathematically regularization involves methods to decide ambiguous cases, for example optical illusions in optics. As an example the usual two-dimensional drawing of a cube can be seen in two different ways, depending on whether the lower left square is viewed as front, or the upper right square is viewed as front. Applying boldface to one or the other squares would favor one way of viewing over the other. Similarly in moral decisions the magnetization field can be viewed as making one decision or the other more likely when there is no other way to decide, as in case 1 above. However at present there is no way to judge whether or not the magnetization field is due to actual evidence or magnetism of the lawyer or propaganda, and so on. Thus more work remains to be done on regularization.

CONCLUSION

The model allows study of many decision-making processes.

REFERENCES

1) Collins, Dennis 2001, Thermodynamic Modelling of Moral Codes (= Moral Codes I), unpublished manuscript, pp.1-42.

2) Collins, Dennis 2003, Moral Codes II, in *Emergy Synthesis 3*, ed. By Mark T. Brown and Eliana Bardi, Center for Environmental Policy, Univ. of Florida, Gainesville, FL, pp.93-102, ISBN 0-0707325-2-X.

3) Jaynes, E.T. 1957, Information Theory and Statistical Mechanics, *Physical Review, Vol* 106, pp.620-630.
4) Godoy, Ricardo 2008, Is Homocide in Puerto Rico High?, *Homicide Studies*, Sage Publications, retrieved on-line2-21-09 estudio_homicidios_pr_eeuu.pdf.

```
ClearAll
n=64;
tuples=Tuples[{1,0},{2,3}];
U=Thumbnail[ArrayPlot[#],25]&/@tuples;
Length[U];
G={{-14,6,0,0},{-6,4,-2,2},{-2,4,0,2},{-2,2,-2,4},
    {-6,4,2,2},{2,2,0,4},{-2,2,2,4},{-2,0,0,6},
    {-6,4,-2,-2},{-6,2,-4,0},{6,2,-2,0},{-2,0,-4,2},
    {2,2,0,0},{2,0,-2,2},{6,0,0,2},{-2,-2,-2,4},
    {-2,4,0,-2},{6,2,-2,0},{2,2,0,0},{2,0,-2,2},
    {6,2,2,0},{14,0,0,2},{2,0,2,2},{2,-2,0,4},
    {-2,2,-2,-4},{-2,0,-4,-2},{2,0,-2,-2},{-6,-2,-4,0},
    {6,0,0,-2},{6,-2,-2,0},{2,-2,0,0},{-6,-4,-2,2},
    {-6,4,2,-2},{2,2,0,0},{6,2,2,0},{6,0,0,2},
    {-6,2,4,0},{2,0,2,2},{-2,0,4,2},{-2,-2,2,4},
    {2,2,0,-4},{2,0,-2,-2},{14,0,0,-2},{6,-2,-2,0},
    {2,0,2,-2},{2,-2,0,0},{6,-2,2,0},{-2,-4,0,2},
    {-2,2,2,-4},{6,0,0,-2},{2,0,2,-2},{2,-2,0,0},
    {-2,0,4,-2},{6,-2,2,0},{-6,-2,4,0},{-6,-4,2,2},
    {-2,0,0,-6},{-2,-2,-2,-4},{2,-2,0,0},{-6,-4,-2,-2},
    {-2,-2,2,-4},{-2,-4,0,-2},{-6,-4,2,-2},{-14,-6,0,0}};
MatrixForm[G];



T=6;b=1/T;h=4;g=5;v=4;
P=Table[Transpose[{{i,U[[i]]},N[Exp[b*(-
G[[i,1]]+h*G[[i,2]]+g*G[[i,3]]+v*G[[i,4]])]/Sum[Exp[b*(-
G[[k,1]]+h*G[[k,2]]+g*G[[k,3]]+v*G[[k,4]])],{k,1,64}]]}}],{i,1,6
4}];MatrixForm[P];
V=Partition[P,4];MatrixForm[V]
Q=Table[Sum[G[[i,m]]*N[Exp[b*(-
G[[i,1]]+h*G[[i,2]]+g*G[[i,3]]+v*G[[i,4]])]/Sum[Exp[b*(-
G[[k,1]]+h*G[[k,2]]+g*G[[k,3]]+v*G[[k,4]])],{k,1,64}]],{i,1,64}],{m,1,
4}
```

1	2	3	4
0.215231	0.0107157	0.0291284	0.00550164
5	6	7	8
0.300379	0.014955	0.15422	0.0291284
9	10	11	12
0.000744566	0.00014063	0.000100766	0.000072202
13	14	15	16
0.00103912	0.000196265	0.000533505	0.000382273
17	18	19	20
0.00202394	0.000100766	0.00103912	0.000196265
21	22	23	24
0.00282463	0.00014063	0.00550164	0.00103912
25	26	27	28
0.0000265616	$5.01684 \cdot 10^{-6}$	0.0000136372	$9.77148 \cdot 10^{-6}$
29	30	31	32
0.0000370697	$7.00157 \cdot 10^{-6}$	0.000072202	0.000051735
33	34	35	36
0.0208714	0.00103912	0.00282463	0.000533505
37	38	39	40
0.110503	0.00550164	0.0567343	0.0107157
41	42	43	44
0.000072202	0.0000136372	$9.77148 \cdot 10^{-6}$	$7.00157 \cdot 10^{-6}$
45	46	47	48
0.000382273	0.000072202	0.000196265	0.00014063
49	50	51	52
0.000744566	0.0000370697	0.000382273	0.000072202
53	54	55	56
0.0039421	0.000196265	0.00767815	0.00145022
57	58	59	60
$9.77148 \cdot 10^{-6}$	$1.84559 \cdot 10^{-6}$	0.000072202	$3.59473 \cdot 10^{-6}$
61	62	63	64
0.000051735	$9.77148 \cdot 10^{-6}$	0.000100766	0.000072202

Detection of Some Malignant 2D Tumors by 1D Continuous Symmetry

By Dennis G. Collins, Dept. of Mathematical Sciences, Univ. of Puerto Rico, Box 9018, Mayaguez, PR 00681-9018

Feb. 15, 2009

ABSTRACT: Malignant tumors show a "tortuous" boundary compared with benign tumors. Based on the Circle Theorem of the author's previous 2008 SIDIM paper "Examples of Measuring Continuous Symmetry," the diameter (distance across) of a two-dimensional plane figure of (perimeter) length one puts a lower bound on its continuous symmetry. If the continuous symmetry of the given tumor figure, normalized to length one, is more than the continuous symmetry of a circle of the same length (perimeter) one, it is shown to be some indication of malignancy. **Keywords:** Continuous symmetry, tumor.

Introduction

This paper explores a possible application of the author's definition of continuous 1D (= one-dimensional) symmetry to medicine, based on the one-dimensional boundary of a tumor drawn in 2D (= two dimensions). Restricted to one-dimension, the continuous symmetry is found by the following steps: 1) normalizing the boundary to length one, 2) calculating the distance between each pair of points on the boundary curve; 3) these distances can be expressed as a function above the unit square; 4) defining a cumulative distribution function as area (within the unit square) versus distance, 5) finding the derivative of the cumulative distribution function as a probability density, going from 0 to the maximum distance (diameter) calculated between two points; 6) finding the entropy of the density, 7) defining the continuous symmetry of the normalized figure as the negative of the entropy and 8) defining the continuous symmetry of the original figure by multiplying the result of the previous step by the normalization constant. Typically in step 3) one of the points can be considered as going downward along the y-axis from 1 to 0 and the other point can be considered as going from left to right from 0 to 1 along the x-axis. In step 6) the entropy is defined as integral of $-p(x) \ln(p(x))$ from 0 to diameter, so that step 7) just leaves off the minus sign.

In general it is very difficult to calculate the continuous symmetry of a figure. The continuous symmetry of some well-known figures is given in the previous paper, "Examples of Measuring Continuous Symmetry," and repeated here:

Figure	Continuous Symmetry
1) straight line (length 1)	.193= ln(2)-1/2
2) L-shape (1/2 length each side)	.476
3) V-shape (60 degrees, ½ length each side)	.741

© 2009 Dennis G. Collins

4) coinciding lines (1/2 length each)	.886=ln(4)-1/2
5) square (1/4 length each side)	1.068
6) equilateral triangle (1/3 length each side)	1.161
7) circle (circumference 1, radius 1/(2*pi)	1.289=2*ln(pi)-1.

It is conjectured that surprisingly the minimum continuous symmetry occurs for a straight line. It might be conjectured that the maximum continuous symmetry occurs for a circle; however this result is not the case as the continuous symmetry can be taken to infinity by folding two lines of length ½ to make four coinciding lines of length ¼, and so on. The results come out as follows:

Figure	Continuous Symmetry
8) 4 coinciding lines (1/4 length each side)	1.579=ln(8)-1/2=3*ln(2)-1/2
9) 8 coinciding lines (1/8 length each side)	2.272=ln(16)-1/2=3*ln(2)-1/2.
And so on	Infinity

However it is observed that one way to get the continuous symmetry to go to infinity is to make the maximum distance across or diameter go to 0. This observation leads to the Circle Theorem:

The continuous symmetry of a figure of perimeter one is at least as large as $ln(1/d)$, where d is the diameter of the figure. This calculation results because the maximum entropy (or minimum continuous symmetry) occurs for the uniform density on the interval $[0,d]$, with height $1/d$. For example four coinciding lines of length ¼ have continuous symmetry 1.579 greater than or equal to $ln(1/(1/4))= ln(4)=1.386 < 1.579 = ln(8)-1/2$. Thus as a figure gets more bunched up or with tortuous boundary, its continuous symmetry goes up.

Thus it may be conjectured that the circle has the most continuous symmetry of a figure of perimeter one without "bunching up." However "bunching up" is more or less equivalent to "tortuous boundary."

An interesting example is to take a circle of perimeter one and "bunch it up" as follows:
Put one pin inside the bottom of a string circle and two pins close together inside the top of the circle; put another pin between the two pins on the top and pull it down until all lines coincide as much as possible. Then there are two lengths of 1/pi=.3183 and two shorter

lengths of (pi/2-1)/pi=.1817. The continuous symmetry comes out 1.390 bigger than that of a circle (1.289), based on the slightly approximated density:

$$f(x) = \begin{array}{ll} 8*(.5+2*.1817-3*x), & \text{if } 0 <= x <= .1366 \\ 8*(1-4*x) & \text{if } .1366 <= x <= .1817 \\ 16*(.3183-x) & \text{if } .1817 <= x <= .3183 = 1/\text{pi}. \end{array}$$

METHOD

The foregoing analysis suggests the following simple method to test for tortuous boundary:

1) Select points around the boundary of the figure following the pattern as closely as possible. Of course this procedure will not allow for fractal boundaries; however it is only necessary to get close to the boundary pattern, since the method is based on inequalities.

2) Compute the sum S of the distances from one point to the next around the boundary. This number S is the approximate perimeter.

3) Compute the diameter D as approximately the maximum distance between any of the points selected.

4) Normalize the perimeter to one, giving the normalized diameter d=D/S.

5) Calculate T= $ln(1/d)$ = $ln(S/D)$, the lower bound on normalized continuous symmetry.

6) If T is greater than the value for the circle (1.289) the pattern is judged as having at least somewhat tortuous boundary.

7) Some number greater than 1.289 (say 1.40) would be a cut-off for malignant tumors, i.e. malignant tumor pattern. The exact number would have to be based on experimental results, and type 1 and type 2 error permitted.

Remark: This method contradicts the expectation of the previous paper where it was guessed that a malignant tumor would have less continuous symmetry. Actually there are opposite tendencies in effect.

RESULTS

For the equilateral triangle, the above T-value (with three vertex points) comes out T=ln(1/(1/3)) = ln(3)=1.0986 < 1.161= "exact," and for the square (with 4 vertex points) T comes out T= ln(1/(sqrt(2)/4))=1.0397 < 1.068 = "exact," so that both the equilateral triangle and square would have T < 1.289 and would be considered non-malignant or non-cancerous, since the continuous symmetry is less than 1.289 and 1.40. The one-dimensional example with the pins has zero enclosed area and the same diameter and perimeter as a circle and so would have T =1.289 < 1.390 = "exact" and would also be cataloged as non-malignant, although somewhat on the border.

For the general case, a simplified version of the algorithm pending patent decision can calculate the T value. Some examples are given. It remains to be seen if this method is valuable in a practical setting. The reader can check the shape by entering the coordinates in a graphing program.

Example 1: (a "triangular" shape based on 9 points that would be classified as malignant) points P= { (4,0),(1,0.8),(4,1),(0.5,2),(0,8),(-0.5,1),(-4,1),(-0.5,0.5),(-4,0)} and back to (4,0) has perimeter S =41.332, diameter d= 8.944, T= ln(41.332/8.944) = 1.530 > 1.40.

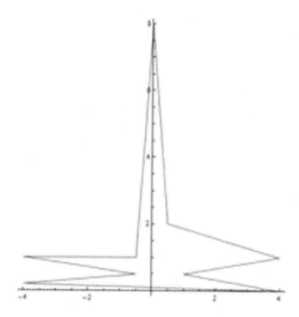

Figure 1. Example 1. T= 1.53.

Examples 2-8 indicate increasing T from a roughly elliptical shape (Example 2) to more star-shaped (Example 8). Added points are underlined, so that the number of points goes from 8 to 10. The number of points by itself doesn't affect results.

Ex. 2:   T= .815 P={(3,0),(2.8,.8),(0,1),(-3.2,.3),(-3.5,-.5),(-2.7,-.8),(-.2,-.9),(2.6,-.7)}

Ex. 3:   T=.960 P={(1,0), same   }

Ex.4:   T=1.166 P={(-1,0), same   }

E5 T=1.20 P={(-1,0),(2.8,.8),(-.3,-.6),(0,1),(-3.2,.3),(-3.5,-.5),(-2.7,-.8),(-.2,-.9),(2.6,-.7)}

Example 6:   T=1.321 P = {(-1,0),(2.8,.8),(-.3,.6),(0,1),(-3.2,.3),(-3.5,-.5),(-2.7,-.8),
                                                     (-.2,-.9),(-1.5,-.5),(2.6,-.7)}

Example 7:   T=1.399 P= {(-2,0), same as Example 6   }

Example 8:   T=1.433 P= {(-2,0),(2.8,.8),(-.8,.6),(0,1),((-3.2,.3),(-3.5,-.5),(-2.7,-.8),
                                                     (-.2,-.9),(-1.5,-.5),(2.6,-.7)}

In Example 8 only the first coordinate of the third point is changed from Example 7. The examples are somewhat exaggerated; however a crinkly boundary all the way around would have the same effect. It must be remembered T does not measure the continuous symmetry itself but only a lower bound on it, so that the actual continuous symmetry is always greater than or equal to T.

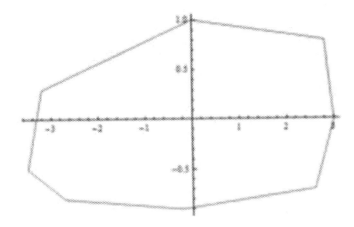

Figure 2. Example 2. T=.815.

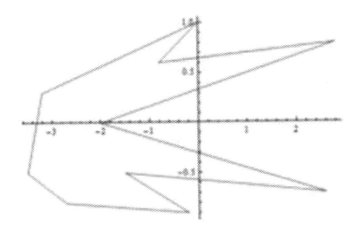

Figure 3. Example 8. T=1.433.

CONCLUSION

A method is devised to detect some 2D malignant tumors by continuous 1D symmetry. Much work and probably refinements are necessary to make the method practical. Also it may be necessary to work with three dimensions.

ACKNOWLEDGMENTS

The author acknowledges Drs. Vito Quaranta and Alissa Weaver, of Vanderbilt University, TN for the Sigma Xi presentation "Integrating Multiscale Data for Cancer Invasion and Metastasis" at University of Puerto Rico, Mayaguez March 15, 2007 and following discussion which raised the question considered here. Since the algorithm is very simple, it may exist in a location unknown to the author at present, apart from the background theory. According to Dr. Weaver, currently judgments of tumor malignancy are made by visual inspection of lab doctors.

REFERENCES

Collins, Dennis G. 2008. Measuring Continuous Symmetry and Positional Entropy, in *General Systems Bulletin, Vol. XXXVII*, ISSS (International Society of Systems Science, pp.20-22.

Collins, Dennis G., Feb. 29, 2008. Examples of Measuring Continuous Symmetry, paper presented at SIDIM 2008, UPR-Carolina, Carolina, Puerto Rico.

POLITICAL SPECTRUM MODELS

Dennis G. Collins (1) and David Scienceman (2)

(1) Dept. of Mathematical Sciences, Univ. of Puerto Rico, Mayaguez, Box 9018 Mayaguez, PR 00681-9018
(2) 17 Cullen Crescent, Warrina Village, Castle Hill, New South Wales, Australia, 2154.

ABSTRACT

This paper studies a threefold interaction in the political setting as a beginning to its study in ecological systems, and emergy maximization. Two *Mathematica* programs are developed. Possible pathology of extending to three interactions is considered.

INTRODUCTION

Most ecological and emergy studies only consider pairwise interactions (Higashi and Burns, 1991). However H.T. Odum's theory of maximum empower derives partly from the mathematical properties of the power function, $f(x) = x^n$, or x to the power n, through the fact that emergy of an interaction is the product of interacting emergies. This fact gives possible survival value to processes that have higher feedback powers $n > 2$. The same survival power may go to processes that involve interaction of three or more <u>different</u> inputs provided they are all sufficiently large (Odum, 1983, p.7). This mathematical fact that, say, $x^3 > x^2$ if $x > 1$ but $x^3 < x^2$ if $0 < x < 1$ is related to Odum's controversy with Prigogine. Thus studying emergy without considering triple and higher products may be somewhat like studying baseball without considering triples or home runs, which have a great impact on the survival and successful competition of baseball terms. A fourfold interaction is suggested by some of the diagrams (Figures 1-4), but only indirectly considered here, in terms of Q multiplied by L, Co, and R.

H.T. Odum published a Left-Right political model with Dr. David Scienceman in 1986 in the *General Systems Bulletin, Vol XXIX* 1985-1986, pp. 23-32 of the International Society of Systems Science. This model had the "nice" property that the political Right had higher transformity than the political Left, and appeared to the right on emergy diagrams. The model also had interesting feedback logic in that a decrease of output for five time periods would trigger a change of power from the Right to the Left and vice versa. The model as given had the property that a slight decrease of output over five time periods caused a change in power from Right to Left, followed by a larger decrease of output for five more time periods, which led to a change of power back to the Right and increased output back to nearly-original levels, resulting in a pulsing of political power. Feedback loops channeled rewards back and forth accordingly. A Scienceman version to extend the logic to include a fuller political spectrum, such as Tough-Tender (Eysenck 1954 and Wilson 1973) or "U-shaped" spectra, was studied by Odum, but never published. The present paper, somewhat comparable to the Thomas Abel presentations at the 4th Emergy Research Conference (such as "Pulsing and Cultural Evolution in China" and "Emergy, Sociocultural Heirarchy, and Cultural Evolution,", Abel 2007) works to update the Odum-Scienceman model.

This paper has a limited goal, namely extending the Scienceman/Odum paper to a larger political spectrum model. This project is done in a series of steps, with the minimal change feasible in the Odum/Scienceman 1986 *BASIC* program.

MODIFICATIONS TO THE 1986 ODUM-SCIENCEMAN PROGRAM

1) non-renewable resources are dropped as an unnecessary complication, although clearly the using up of non-renewable resources, such as Alaskan oil, can play a big role in distinguishing parties.

2) A co-operative party "Co" is added to the left L and Right R (Remark: "C" is a reserved symbol in *Mathrmatica* and thus can't be used.). This grouping is called "Choices" in the Scienceman/Odum study of the political spectrum.

3) A threefold feedback product k*L*Co*R is added to the assets Q. This kind of a term is a missing ingredient in the Scienceman/Odum study [Odum and Scienceman, 1985-1986], although [Odum and Scienceman, 2005] do include it in their Marxian models. Maximum feedback power (the "Triple Helix effect" [Burhoe, 1973, Odum, 1977, Lewontin, 2000]) is attained when all three of the factors are operating. The Co party can operate as a "Silent Majority" even if it has no logical status (i.e. representation in the government).

4) A global stress variable "S" is added to the local (5-period) stress variable "X" of the 1986 paper. The global stress S is increased by 1 after the local stress X has cycled three times.

5) Logic statements are added with a binary (0 or 1) variable "U" to specify when the Co party has effect. In the 1986 Scienceman/Odum model "Z" specifies when the left L has effect and "W" specifies when the right R has power. L and R are mutually exclusive in the 1986 paper, that is, they cannot both be activated (1) at the same time.

An optional addition is "Tough-Tender" feedback, which is included in the unpublished Scienceman/Odum political spectrum study of 1986. These feedback terms can eliminate switching and oscillation among political parties.

6) Terms of the form (L-10*R)*U can switch resources from Left to Right and vice versa as required to help maintain traditional ratios of storage assets (in this case R has a higher transformity 10 times L). If L= 10*R or if U is off (=0), there is no switching. In previous models the transformity of R over L (the specific value 10 in the term above) comes from "self-organization" of the feedback constants versus setting it "by hand."

7) "Sticky" transfers can be included in which some of the resources switched between L and R "stick" to the Co party along the way through a feedback term to the Co party, k*Abs[L-10*R]*U. This term could be viewed as a form of political corruption.

THE CO-OPERATIVE PARTY

The Co-operative party (or Choices Party) originates in studies of the political spectrum by Eysenck (1954) and Wilson (1973). These studies detect a classification axis called "Tough-Tender" more or less orthogonal to the standard Left-Right axis. The "Tender" grouping along this axis represents tolerant people who would reather make love than war, and tend to come together, whereas the "Tough" grouping represents strict people who would rather make war than love. For example the English monarch Charles I cannot accommodate Cromwells's Parliament because "There is no power above the King." As a consequence, the political spectrum takes on a "U" shape, with the Co party at the bottom of the "U," the Left at the top left of the "U," and the Right at the top right of the "U." Then the "Tough-Tender" axis is vertical with the "Tough" people at the top (splitting into Left and Right) and the "Tender" people Co at the bottom.

THE "TRIPLE HELIX" FEEDBACK TERM

Why didn't H.T. Odum publish in 1986—it could be because of not including a "Triple Helix" feedback term k*L*Co*R in the production assets variable Q. Actually the double product terms, such as L*R, added to the Q variable in the 1986 paper are already more or less "off the charts" as far as mathematical nonlinearities are concerned, since they do not exactly fit into the quadratic "community matrix" pattern (Higashi and Burns 1991), which could include L*R (symbiotic) terms [Odum, 1983, p.217] in the L or R storage but not in the Q storage. Nonetheless the two-factor symbiotic products L*R can be extended into a three-factor term L*Co*R symbiotic term. This threefold idea was discussed by Ralph Wendell Burhoe (1973), H.T.Odum (1977)and Lewontin (2000). Translated into music, it says music is a combination of three factors: the composer (say DNA or genotype), the performer (say specific organism or phenotype), and the listener (or environment). Without all three, the music falls short. The specific three factors change from setting to setting.

GLOBAL STRESS VARIABLE

To start putting into the program the logic to determine when the Co-operative party Co takes power, we introduced a global stress variable S. This variable measures when production falls over repeated cycles of the local stress variable X. In this program the global stress S is increased by 1 every time the local stress X goes through 3 cycles of five downturns of production, i.e. in this program 15 steps of decreasing production, not necessarily in order (may be spread out in time, for example over 25 years). Here people are willing to try something else if things keep on going downhill and the left L and right R parties merely cycle back and forth. In this program, the Co-operative party turns on by changing its binary indicator U from 0 to 1 (roughly gains power) after S reaches 3, and turns off by changing U back to 0 (roughly loses power) after S reaches 4, when it (the global stress variable S) is set back to 0. Many other ways to define the change logic could be devised, for example based on decline of production or other variables., such as corruption.

The Co Party Indicator "U" and Classification of Political Spectrum

The Co Party indicator "U" being 1 also represents the "Tender" setting being on, versus U being 0 standing for the "Tough" setting being on. The three binary indicators: Z for the Left L, U for the Co-operative party, and W for the Right R permit an eight-fold classification of political parties, i.e. the political spectrum as follows: ZUW =

1) 100 the Left
2) 110 the Center-Left
3) 010 the Center or Co-operative Party
4) 011 the Center-Right
5) 001 the Right

Since Left and Right are mutually exclusive in the 1986 Odum/Scienceman program,

6) 101 the Gridlock Party cannot occur, since Z=1 and W=1 at the same time.

Also 7) 111 the "National Unity/Chaos" Party with everyone in control canot occur.

Then 8) 000 the "Anarchy Party" with no one in control, or since in some sense U = 0 stands for "Tough" versus "Tender, "tough rule with neither left nor right representation 000 could code for dictatorship.

The first five code values can be considered as points along a "U" shape, with upper left of the "U" being 100, further down being 110, bottom being 010, along the upward right slope of the "U" being 011, and the upper left point being 001.

Actually the original Eysenck-Wilson U-shape [Eysenck, 1954, Wilson, 1973] is more strict or extreme with (in my notation) 100 coding Communist, 110 coding Left, 010 coding Center, 011 coding Right, and 001 coding Fascist.

Also (communication by Scienceman), since the bottom 010 of the U shape, representing the "Tender" co-ordinate value 1 is supposed to be more optimal versus the Communist and Fascist parties with "Tender" co-ordinate 0; in Europe (Wilson and Caldwell 1988), the "U" shape is graphed upside down or inverted. This upside-down U has the Fascist and Communist parties on the bottom. One of us, Scienceman, asked if the top of the inverted U somehow represents maximum empower.

Anyway, with the added feedback terms, stress variables, and logic and a few modified feedback strengths in place, we can let the modified Odum/Scienceman program run. The result is the program denoted sccog13, listed later in Program 1. A main feature of this program is that, although all five parties gain power at various points, they all eventually build up their coffers at the expense of the general assets Q, causing Q to go down and the party to go out of power. This pattern is the "Power corrupts and absolute power corrupts absolutely." idea. However a second feature seems to be that the three parties self-organize to about the same strength, i.e. to the same transformity. More of this possibility later.

TOUGH-TENDER TRANSFERS

Very little of the above coding is contained in the original Scienceman/Odum model, which instead views the Co grouping as managing transfers of resources from Left to Right and vice versa, so as to maintain status-quo of transformity between Right and Left, in their case a transformity of 10 to 1 for Right over Left. (This property may be another reason H.T. Odum didn't publish.) A term -k*(L - 10*R)*U is added as feedback to the Left, and a corresponding term k*(L-10*R)*U as feedback to the Right. If say L becomes more powerful at 200 and R stays at 10, representing a 200/10 = 20 to 1 transformity, then -k*(L-10*R) *U represents a deduction -k*(200-100)*1 = 100k from the Left, which is added to the Right if the "Tender" variable U=1, or Co has effect. On the other hand if R becomes more powerful from 10 to 20, and L stays at 100, representing a transformity of 100/20=5, then -k*(L-10*R)*U stands for an addition of amount -k*(100-200)*1 = 100k to the Left, and its subtraction from the Right, again if U =1. These terms tend to keep the whole system at equilibrium and thus prevent oscillations.

STICKY TRANSFERS

Finally it may be likely that the Co Party charges for its transfer service, so that some of the resources transferred "stick" to the Co coffers, which thus increase. This process can be modeled by a term k1*Abs[L-10*R]*U fed back to the Co storage, together with making the other transfers less. The absolute value Abs function means that Co collects regardless of which way the transfer goes. In any case, this term is consistent with H.T. Odum's policy of attaching losses to every process. In the long term, the Co party may become simply another pig at the feeding trough, increasing its storage at the expence of the general assets Q. However in the program as developed, it did stop political oscillation, leaving the Center-Left Party in control.

POSSIBLE PATHOLOGY

There is the possibility of a pathological growth of the Co party.

This paper started with the execution of Saddam Hussein in 2007 and its similarity to the death of Mithradates VI (Eupater), who led several rebellions against Rome, prior to his death in 63 b.c. Although Saddam Hussein didn't claim to speak the 22 languages of all his provinces in Asia Minor as apparently did Mithradates VI, there is still some sense that he may have done a better job ruling Iraq than the democratic process set up there since the Iraq war of 2003. The Founding Fathers of the U.S. republic had the idea that democracy could degenerate into dictatorship; but "What are the specific feedbacks and instabilities involved?"—could the same instabilities beset the U.S. as caused the Roman republic to falter within 20 years of the end of Mithradates VI. Julius Caesar's unlimited dictatorship ended with his assassination in 44 BCE but with dictatorial rule continuing in Rome, causing a loss of confidence that democracy is best.

Some versions of the model lead to unbounded growth of the Co party, which can form a rough medical model of cancer growth (Figure 3).

Recently Mayor Bloomberg did away with term limits in New York City, followed by Chavez in Venezuela and probably Putin in Russia. Articles have mentioned the possibility of doing away with term limits in the United States, so that Obama could stay in power, or with the domestic birth requirement of the Constitution, so that Arnold Shwarzennager could take over. These developments could lead to a return to the paternal model of politics, such as practiced in North Korea. Interestingly the U.S. balance of power structure did work to keep Franklin Roosevelt from taking over the Supreme Court in the 1930's. Some political units may have difficulty keeping the volume of threefold interaction sufficiently high to maintain the benefits, as mentioned in the controversy versus Prigogine. For example in Puerto Rico the legislature was reduced to one chamber to save money, but the recent Governor was forced to resign because of supposed power abuse.

RESULTS

The two main programs (sccog13 and sccog17) modifying the 1986 Odum/Scienceman program seemed to be somewhat robust, perhaps thanks to the original protgram, which worked as planned. The first program (sccog13) shows a moderate decrease of political oscillation, but with most parties running out of steam, or becoming corrupt. The change of political party can be seen as positive in getting out of power parties that cause a precipitous decline of general assets. Co was set at 100 to start but seemed to self-organize to around 25, midway between initial values of 50 for L and 5 for R.

The second program, following more the "Tough-Tender" idea, did stabilize the change of power, but with more rigidity in that the Right-Left transformity is set ahead of time in the program (at 10) instead of being allowed to "self-organize."

Thinking in terms of maximum empower, a first attempt to maximize empower might be to maximize the feedback term $k*L*Co*R$ to the general assets Q. Given transformities of a,b, c for factors A,B,C the problem max $A*B*C$ subject to $a*A +b*B+c*C$ = constant emergy Em has an "equipartition of emergy "solution $a*A = b*B=c*C=Em/3$. In the case of given transformities a=1, b=2, c=10, this result would come out something like $50*1 = 25*2= 10*5$ if Em = 150. The product would be $A*B*C=50*25*5=6,250$. On the other hand if A, B and C all self-organized to the value 25, with common transformity 2, the product would be $25^3 = 15,625$ with much greater feedback to Q. The program sccog13 ends with L=40.6, Co=25.2, and R=25.8, i.e. with values tending together to the same value. The question remains if this result is a general tendency for this type of program, and if so, whether or not it is a good outcome. Some might regard the egalitarian tendency as good, and others the common transformity as a loss of diversity. The program sccog17 ends with L=324.5, Co=110.9, and R=16.2, i.e. with an R to L transformity of about 20. Thus the objective of keeping the R to L transformity at 10 was only partially met, and the question arises where the value 20 comes from. Also in the sccog13 program in the graph of power P, the power rises to above 4500 then trails off, whereas in the sccog17 program P rises above 4500, trails off, but then returns to a value above 4500 after the Center-Left party takes over at time 51. Somewhat the same pattern recurs for the general assets storage Q. Of course in these programs, it is the decline of Q that fuels the change of party, so that with no decline of Q, there is no change of party after time 51 in program sccog17. A question is the extent to which the "Triple Helix" (i.e. threefold) feedback actually occurs, and why the Scienceman/Odum transfer terms (being the first two terms in the DL line, the first term in the DCo line and the first two terms in the DR line of sccog17) increased P more than having equalized storages, i.e. common transformity (sccog13). The idea of maximum empower would seem to favor the Sceinceman/Odum transfer terms coming into play.

CONCLUSION

The business cycle was generally regarded as harmful, and steps were taken to prevent it through the introduction of the Federal Reserve System in the United States. The Federal Reserve can be considered to have higher transformity than the business and banking world because it has feedback loops to control these entities. On the other hand, the ability to change power from one party to another is at present considered a positive characteristic of democracy. The constant change of party can be considered harmful through lack of continuity, and party bickering can be considered one source of a tendency toward dictatorship that the Founding Fathers worried about. Also campaigning and elections represent a drain on resources. However the question arises: if a mechanism is put in place to prevent change of party, is there still democracy. The Roman Republic declined through the introduction of a Triumverate, with apparently Caesar as a control on the wealth of Crassus and the popularism of Pompey. In the models of this paper, the Co Party is designed with a transformity of 2 intermediate between the Left 1and Right 10. However if the Co Party functions as a feedback control on Left and Right, the question comes up, "Does it actually have a higher transformity than both Left and Right?", and, in effect, is this what it means to have a dictator status. Thus it remains to extend the models studied here to include the possibility of dictatorship and then to investigate what types of instability would lead to it. Mathematically these models are considered "switched systems," since the dynamics depend on the logic of which party is in power [Shorten et al., 2007]. Perhaps a fourth variable to the right of ZUW is required, although as mentioned before dictatorship could be coded 000 or possibly 001 (Fascist).

A goal is to extend these models to the "Triple Helix "of Environment-Phenotype-Genotype with longer turnover time going to the right (cf. Andrade); such a procedure is suggested by (Odum, c.1999).

ACKNOWLEDGMENTS

Remark: As it remains to be seen how much I'll be able to consult with Dr. Scienceman, I'm (Collins) writing this by myself at present, although most of the ideas appear in Scienceman/Odum works and papers he (Scienceman) has supplied to me. Many thanks are also due to an anonymous reviewer.

REFERENCES

Abel, Thomas, 2007. Pulsing and Cultural Evolution in China, Chapter 37-1 in *Emergy Synthesis 4*, The Center for Environmental Policy, Gainesville, FL.

Andrade, Eugenio, The Interrelations between Genotype/Phenotype/Environment: A Semiotic Contribution to the Evo:Devo Debate, *Entropy*, retrieved from Internet http://www.library.utoronto/ca.

Burhoe, Ralph Wendell, The concept of God and Soul in a Scientific View of Human Culture, *Zygon*, vol.8,No.3-4, Sept.-Dec.1973, pp.434-436.

Eysenck, H.J., 1954. *The Psychology of Politics*, Routledge and Kegan Paul.

Higashi, M. and T.P.Burns, 1991. *Theoretical studies of ecosystems*, Cambridge University Press, Cambridge, p.26.

Lakoff, George, 1996, 2002. *Moral Politics*, University of Chicago Press, Chicago, IL.

Lewontin, Richard, 2000. *The Triple Helix*, Harvard University Press, Cambridge, MA.

Odum, H.T. (c.1999) and Jan Sendzimir, Limits to Memory in Ecosystems and Society, (unpublished? manuscript). The last diagram contains several triple products, perhaps with misprints): P1=R*Bs*Ns, P2= Bs*Ns*Ns, P3= R*Bs*B*N, P4= B*N*N.

Odum, H.T., The Ecosystem, Energy, and Human Values, *Zygon*, 12(2), pp.109-133.

Odum, H.T., 1983. *Systems Ecology*, John Wiley & Sons, New York., p.7, p.217.

Odum, H.T. and David Scienceman, 1985-1986. Commonalities Between Heirarchies of Ecosystems and Political Institutions, *General Systems, Yearbook of the Society for General Systems Research*, vol. XXIX, pp.23-32.

Odum, H.T. and David Scienceman, 2005. An Energy Systems View of Karl Marx's Concepts of Production and Labor Value, *Emergy Synthesis 3*, The Center for Environmental Policy, Univ. of Florida, Gainesville, FL, pp.17-44.

Shorten, Robert, Fabian Wirth, Oliver Mason, Kai Wulff, and Christopher King, 2007. Stability Criteria for Switched and Hybrid Systems, *SIAM Review*, vol. 44, No.4 (Dec.2007), pp. 545-592.

Wilson, Glenn, 1973. The Liberal Extremists, *New Society*, vol.26, pp.263-264.

Wilson, Glenn and Florence Caldwell, 1988. Social Attitudes and Voting Intentions of Members of the European Parliament, *Pers. Individ. Diff.*, Vol.9, No.1, pp.147-153.

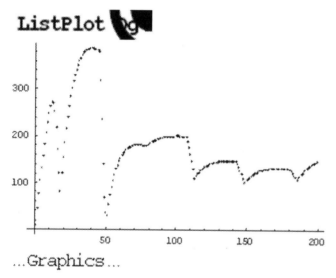

Graph 1. Assets Q versus time for first model (sccog13).

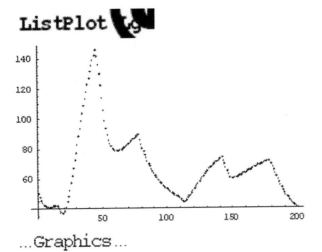

Graph 2. Left storage versus time for first model (sccog13).

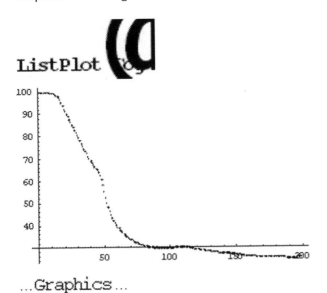

Graph 3. Co-operative storage versus time for first model (sccog13).

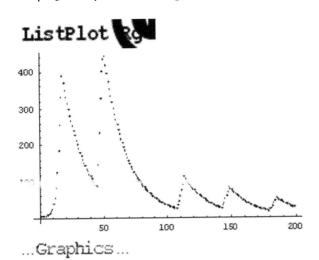

Graph 4. Right storage versus time for first model (sccog13).

ListPlot

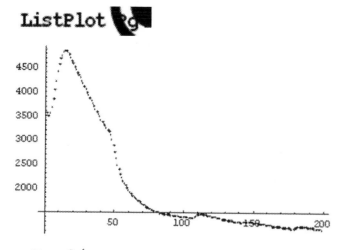

...Graphics...

Graph 5. Output production versus time for first model (sccog13).

```
Co=100; Q=10; L=50; R=5; J=50; k0=.01; k1=.00001;
k2=.01; k3=.0001; k4=.01; k5=.00001; k6=10^(-6);
k7=.01; k8=.02; k9=.002; T=1; inc=1; Lo=10;
Ro=2; S1=.05; S2=.05; S3=.05; L1=.001; L2=.0002;
To=2; Qo=50; X=0; S=0; U=0; W=0; Z=0; Qg={};
Lg=0; Cog={}; Rg={}; Pg={};
While[T/To < 100, Qg=Append[Qg, Q]; Cog=Append[Cog,Co];
Lg=Append[Lg, L]; Rg=Append[Rg, R];
Y= J/(1+k0*L*R);
If[DQ<0, X= X+1]; If[X>5, {Z=Z+1, S=S+1}];
If[X>5, X= 0];
If[Z==2, Z=0]; If[Z==1, W= 0]; If[Z==0, W=1];
If[S==3, U=1]; If[S==4, {Z=0} And {W=0}];
If[S==5, {Z=0} And {W=1}]; If[S>5, {U=0} And {S=0}];
DQ= k3*L*R*Co*Y - S1*Q - k7*Q - k4*Q - Z*k8*Q-
W*k9*R*Q - S*k3*Q;
DL= k7*Q + Z*k8*Q - k6*L*R*Co - k5*L*R*Y - S2*L;
DCo= k1*Q+S*K3*Q -k6*L*R*Co - k3*Co;
DR= L1*Q - S3*R + W*k9*R*Q - k6*L*R*Co - k6*L*R*Y;
P= k0*Y*L*R*Co; Pg= Append[Pg, P];
L= L + DL*inc; If[L<0, L=0];
Q= Q + DQ*inc;
R= R+DR*inc; If[R<0, R=0];
Co= Co + DCo*inc; If[Co<0, Co= 0]
Print[T," ",Q," ",DQ," ",X,S," ",Z,U,W,
""," L," ",Co," ",R," ",L*R*Co];
T= T + inc]
```

Program 1. First program sccog13.

```
1 44.9143 34.9143   00 001
  47.5393  99.9651  4.83143   22960.3
2 76.158 31.2437   00 001
  45.5537  99.9326  5.04233   22954.2
3 104.87 28.712  00 001  43.9798  99.9004  5.60796  24639.1
4 131.893 27.0231   00 001
  42.7693  99.8668  6.58045   28106.6
5 157.767 25.8743   00 001
  41.8847  99.8301  8.08736   33816.1
6 182.71 24.9425   00 001
  41.2958  99.7878  10.3549   42670.7
7 206.574 23.8638   00 001
  40.9749  99.737  13.7571   56221.1
8 228.785 22.2109   00 001
  40.8932  99.6729  18.899   77031.2
9 248.249 19.4642   00 001
  41.0151  99.5882  26.749   109259.
10 263.225 14.9763   00 001
  41.2917  99.4714  38.8267   159475.
11 271.175 7.9495   00 001
  41.6528  99.3046  57.4248   237527.
12 268.708 - 2.46676   00 001
   41.9964  99.0599  85.7267   356637.
13 252.019 - 16.6891   10 001
   42.1784  98.696  127.418   530423.
14 218.6 - 33.4185   20 001
  42.0102  98.1583  184.988   762825.
15 170.877 - 47.7233   30 001
  41.2835  97.3878  256.066  $1.02952\,'\,10^6$
16 119.642 - 51.2355   40 001
  39.849  96.3503  329.911  $1.26668\,'\,10^6$
17 80.1359 - 39.5056   50 001
  37.7367  95.0751  391.206  $1.40358\,'\,10^6$
18 120.133 39.9975   01 100
  36.8007  93.6709  370.317  $1.27654\,'\,10^6$
```

```
19 155.804 35.6702   01 100
   37.2385   92.3982   350.64   1.20647' 10⁶
20 187.614 31.8101   01 100
   38.7946   91.1996   332.052  1.17482' 10⁶
21 215.958 28.3446   01 100
   41.2588   90.0363   314.457  1.16814' 10⁶
22 241.174 25.216    01 100
   44.4569   88.8829   297.777  1.17665' 10⁶
23 263.553 22.3785   01 100
   48.243    87.7239   281.948  1.19322' 10⁶
24 283.348 19.7957   01 100
   52.4945   86.5509   266.916  1.21272' 10⁶
```

Table 1. Some output of first program sccog13.

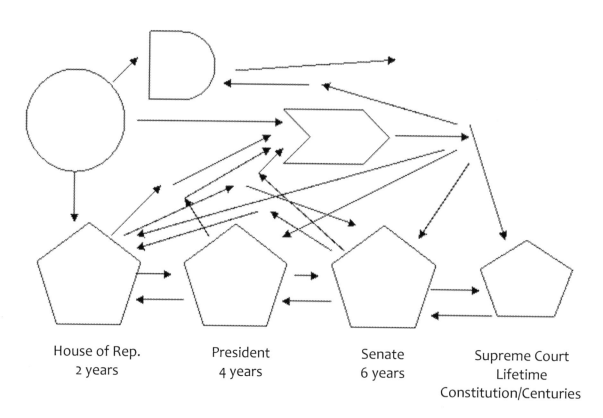

House of Rep.	President	Senate	Supreme Court
2 years	4 years	6 years	Lifetime
			Constitution/Centuries

Figure 1. Interaction of United States Government—Balance of powers. Increasing term of service indicates higher transformity. One of us, Scienceman, suggests the U.S. President plays the role of the Co-operative Party. Fourfold interaction could include Supreme Court/Constitution.

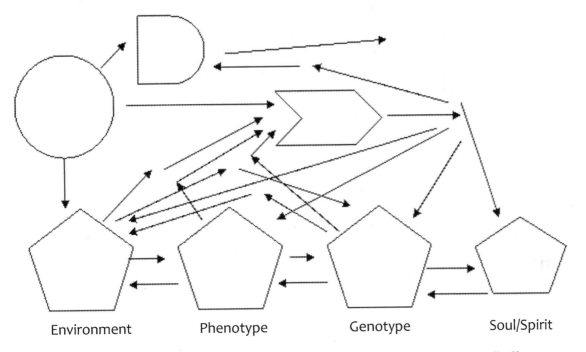

Environment   Phenotype   Genotype   Soul/Spirit

Figure 2. Threefold interaction of programs of this paper. Possible "Triple Helix" effect.

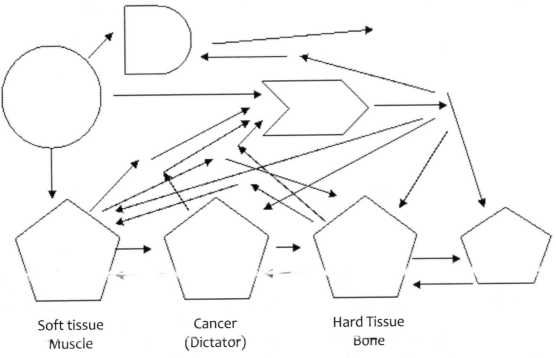

Soft tissue   Cancer   Hard Tissue
Muscle   (Dictator)   Bone

Figure 3. Possible pathology of threefold interaction.

Q P

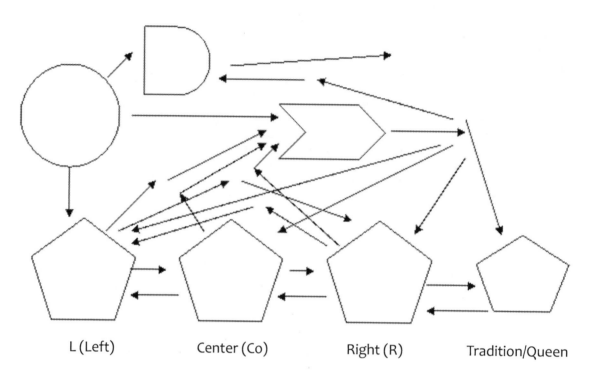

L (Left)   Center (Co)   Right (R)   Tradition/Queen

Figure 4. Political Spectrum Model

110 (Center Left)   010 (Center)   011 (Center Right)

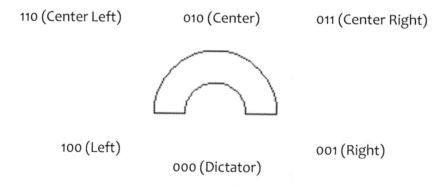

100 (Left)   001 (Right)

000 (Dictator)

Figure 5. Political Spectrum according to this paper (cf. Wilson /Caldwell 1988). Lakoff (2002) argues that U.S. politics combines the two (Left-Right horizontal and Tender-Tough vertical) coordinate axes into one coordinate from Left—Nurturing Parent to Right —Strict Father.

Sccog 13

"Year"	Political Party Change	Triple Product	Value L*	Co*	R at takeover
1	001 Right	22,960	47	99	4
18	100 Left	1,276,540	36	93	370
45	001 Right	1,332,180	141	63	147

51	110 Center-Left	2,068,190	96	50	417
78	010 Center	259,367	85	30	98
107	001 Center-Right	48,667	47	30	33
113	100 Left	143,081	44	29	107
142	001 Right	63,865	71	27	32
148	001 Left	126,684	60	26	78
179	011 Center-Right	38,607	69	25	21
185	010 Center	73,260	56	25	51
199		26,435 40 25 25			

Table 2. Political Party Changes in first program sscog13.

Sccog 17

"Year"	Political Party Change	Triple Product	Value L*	Co*	R at takeover
1	001 Right	22,960	47	99	4
18	100 Left	1,267,540	36	93	370
45	001 Right	1,332,180	141	63	147
51	110 Center-Left	2,844,490	120	62	374
199		521,343	304	108	15

Table 3. Political Party Changes in second program sscog 17
(Center-Left Party keeps power after "year" 51.).

Dennis G. Collins and David Scienceman
Jan. 31, 2007
Revised March 23, 2009
Revised Sept. 5, 2009

Dennis G. Collins, Curriculum Vitae

March 8, 2011
7108 Grand Blvd.
Hobart, IN 46342-6628

1519 S. State Rd 118, Apt. 2
Winamac, IN 46996-8550

Urb. Mayaguez Terrace
6009 Calle R. Martinez Torres
Mayaguez, PR 00682-6630

Phone: 787-951-4208, 787-834-3680, 787-612-3680, 517-256-0044
e-mail: d_collins_pr@hotmail.com

ACTIVITIES SINCE RETIREMENT AT END OF DECADE, Dec. 31, 2009

General Systems Bulletin Vol. XXXIX 2010 includes **Moral Codes III: Spin and Regularization in Judgement, pp.16-18.**

Publication of **Emergy Synthesis 5, Center for Environmental Policy, Univ. of Florida, Gainesville, FL,** Dec. 2009 includes 2 papers:
Political Spectrum Models, co-authored with David Scienceman, pp. 23-32.
Architecture Case Study in Transformity Factorization, pp. 599-603.

U.S. Patent 7,873,220 B1 Algorithm to Measure Symmetry and Positional Entropy of a Data Set, issued Jan. 18, 2011.

Feb.25, 2011 talk Algorithms to measure discrete and continuous symmetry: Symmetry Analysis of Howe's Patterns, SIDIM XXVI Conference, Humacao, PR.

Nov. 6, 2010 talk Continuous symmetry of the wedge and other shapes (ABSTRACT 1064-51-167) 1064th Meeting of American Math Society (AMS), U. of Notre Dame, South Bend, IN.

Oct. 16, 2010 talk Detection of some malignant 2D tumors by 1D continuous symmetry, Indiana Section meeting of Math Association of America (MAA) Purdue North Central, Westville, IN.

Feb. 26, 2010 talk Toward a Mathematical Origin of Species, presented at SIDIM XXV Conference, Mayaguez, PR.

Jan. 15, 2010 talk Some Continuous Empower2 Models at 6th Emergy Synthesis Conference, Gainesville, FL.

Feb. 5, 2011 and Feb. 6, 2010 Judge at SESO High School Science Fair, Mayaguez, PR.

Still serving as Treasurer of local #511 Chapter of Sigma Xi research society.
Listed Marquis Who's Who in World, Who's Who in America, Who's Who in Science and Engineering.

Dennis G. Collins, Curriculum Vitae   Oct. 23, 2009

Urb. Mayaguez Terrace
6009 Calle R. Martinez Torres
Mayaguez, PR 00682-6630
Phone: 787-832-4040 Ext. 3285
        787-834-3680 (Home) 787-612-3680 (Cell)

Birthdate: 6-26-44
Height: 5'9.5" Weight: 190 lbs.
U.S. Citizen Married
e-mail: d_collins_pr@hotmail.com

ACADEMIC RECORD

Graduate: 1970-1975. Ph.D. (Math) Illinois Institute of Technology (IIT), Chicago, IL.
        GPA 3.41/4.00 including advanced applied courses in six other departments.
        1-1970. M.S. at IIT, Chicago.
        1966-1967. National Science Foundation Fellowship at Yale Univ., New Haven, CT
Undergraduate: 1962-1966. Valedictorian; GPA 4.00/4.00, class of 200, Merrillville
        High School, Merrillville, IN.

ACADEMIC EXPERIENCE (list of courses offered available upon request)

1990- .   Professor, Univ. of Puerto Rico, Mayaguez Campus, Mayaguez, PR;
2003-2004. Sabbatical—at UPR-Mayaguez, PR to study entropy and synergy.
1996-1997. Sabbatical—Visiting Scholar, Michigan State Univ., E. Lansing, MI.
1986-1990. Assoc. Prof. UPR-Mayaguez;
1988-1989. Sabbatical—Visiting Associate Prof. and Visiting Scholar at Michigan State Univ.,
        E. Lansing, MI and Univ. of New Orleans, New Orleans, LA.
1982-1986. Assistant Prof., UPR-Mayaguez;
1979-1982. Assistant Prof., Dept. of Math and Computer Science, Valparaiso Univ., Valparaiso,
        IN.
1976-1979. Instructor, Univ. of New Orleans, New Orleans, LA.
1975-1976. Instructor (temporary one-year), Michigan State Univ., East Lansing, MI.
1970-1975. One-half and one-third Teaching Assistant at IIT, Chicago.

ABSTRACTS/PAPERS

2009 Examples of Measuring Continuous Symmetry, published in 2009 *General Systems Bulletin*, Vol. XXXVIII, International Society for Systems Science, pp.12-14.

2008 Measuring Continuous Symmetry and Positional Entropy, presented as poster July 9, 2007 at MAXENT 2007, Sarasota Springs, NY; published in 2008 *General Systems Bulletin, Vol. XXXVII*, International Society for Systems Science (ISSS), pp.20-22.

2007 An Algorithm to Measure the Symmetry and Positional Entropy of n-Points, presented at annual meeting of American Mathematical Society (AMS), Jan. 6, 2007 AMS Abstract #1023-52-1900; published in 2007 *General Systems Bulletin, Vol. XXXVI*, International Society for Systems Science (ISSS), pp.16-21. Curriculum Vitae for Dennis G. Collins, p.2.

2006 "Tropical" Emergy and (Dis-) Order, in *General Systems Bulletin, Vol.XXXV*, ISSS, pp.17-22; revised version in *Emergy Synthesis 4*, ed. By Mark T. Brown and Eliana Bardi, Center for Environmental Policy, Univ. of Florida, Gainesville, FL, Chapter 14, pp.1-6. ISBN: 978-0-9707325-3-8.

2005 Moral Codes II, in *Emergy Synthesis 3*, ed. By Mark T. Brown and Eliana Bardi, Center for Environmental Policy, Univ. of Florida, Gainesville, FL, pp.93-102, ISBN 0-0707325-2-X.

2005 Algorithm for Minimum Laterally-Adiabatically Reduced Fisher Information, in *Bayesian Inference and Maximum Entropy in Science and Engineering, 25th Int. Workshop, San Jose, California*, ed. by Kevin Knuth, Ali Abbas, et al., AIP (American Institute of Physics), Conference Proceedings #803, Melville, NY, pp.345-354, ISBN 0-7354-0292-2.

2004 Divvy Economies Based on (An Abstract) Temperature, in *Bayesian Inference and Maximum Entropy Methods in Science and Engineering, 23rd Int. Workshop, Jackson Hole, Wyoming*, ed. by Gary Erickson and Yuxiang Zhai, AIP (American Inst. of Physics) Conf. Proc. #707, Melville, NY, pp.67-74, ISBN 0-7354-0182-9.

2003 On the Rationale of the Transformity Method, in *Emergy Synthesis 2*, ed. by Mark T. Brown, Center for Environmental Policy, Univ. of Florida, Gainesville, FL, pp.171-184. ISBN 0-9707325-1-1.

2003 Transformities from Ecosystem Energy Webs with the Eigenvalue Method, with H.T.Odum, in immediately above, pp.203-220.

2001 Temperature-Based Ascendancy Derived from a Cost or Reward Function, in *Bayesian Inference and Maximum Entropy Methods in Science and Engineering, 19th Int. Workshop, Boise, Idaho*, ed. by Joshua Rychert, Gary Erickson, and C.Ray Smith, AIP (American Insti. of Physics) Conference Proceedings #567, Melville, NY, pp.155-176, ISBN 0-7354-003-2.

2000 Calculating Transformities with an Eigenvector Method, with H.T.Odum, in *Emergy Synthesis*, ed. by Mark T. Brown, Center for Environmental Policy, Univ. of Florida, Gainesville, FL, pp.266-281, ISBN 0-9707325-0-3.

2000 IPO's and the Vapor Pressure Curve, in *Proceedings of the World Congress of the Systems Sciences and ISSS 2000 the 44th Annual Conference of the Int. Society for the*

Systems Sciences, July 16-22, Toronto, Canada, ed. by Janet K. Allen and Jennifer Wilby, ISBN 0-9664183-5-2.

Curriculum Vitae for Dennis G. Collins, p.3

1998 Business-Generalized Thermodynamics, in *Proceedings of the 42nd Annual Conference of the International Society for the Systems Sciences*, July 19-24, 1998, Atlanta, GA, ed. by Janet K. Allen and Jennifer Wilby, ISBN 0-9664183-0-1.

1992 Measurement of Temperature in Economic Systems Considered as Thermodynamic Models, *Energy Economics, Vol. 14*: 1 Jan 1992, pp.73-78.

1991 Canonical Forms for the Digraph Isomorphism Problem, in *Graph Theory, Combinatorics and Applications*, Proceedings of the 6th Quadrenial Int. Conf. on The Theory and Application of Graphs, held in Kalamazoo, MI 1998, published By John Wiley, pp.271-286 of Vol.1 AMS Abstract #88T-05-3.

1991 A Thermodynamic Version of the Phillips Curve, in *Advances in Information Systems Research*, Proc. Of the 5th Int. Conf. on Systems Research, Informatics, And Cybernetics, held in Baden-Baden, Germany, Aug.6-12, 1990, published by Int. Institute for Advanced Studies in Systems Research and Cybernetics, Windsor, Canada, pp.459-464.

1990 A Thermodynamic Setting for the Phillips Curve, in *Mathematical and Computer Modelling, Vol. 14*, Proc. Of the 7th Int. Conf. on Math and Computer Modelling, held in Chicago, IL, Aug. 1989, published by Pergamon Press, Oxford, pp.1183-1188.

ABSTRACTS/TALKS/POSTERS

2009 March 19. Moral Codes III: Spin and Regularization, poster presented in Brain Awareness Week UPRM 2009, Mayaguez, PR.

2009 March 6. Detection of Some Malignant 2D Tumors by 1D Continuous Symmetry, Presented at SIDIM-24, UPR-Rio Piedras, PR.

2008 July 14. Architecture Case Study in Transformity Factorization, paper presented At 52nd Annual Meeting of the International Society of Systems Science (ISSS), Madison, WI; also presented at *5th Biennial Emergy Research Conference*, Univ. of Florida, Gainesville, Jan.31, 2008. Also gave poster with Dr. David Scienceman, Political Spectrum Models, at this (Florida) conference.

2008 Feb.29. Examples of Measuring Continuous Symmetry, presented at SIDIM-23 Conference, UPR-Carolina, PR, submitted to ISSS.

2007 July 11. Thermodynamic Modeling for Social Science Problems, presented at *INFORMS Int. Puerto Rico 2007 Conf., Wyndham Rio Mar Beach Resort and Spa*, PR.

Curriculum Vitae for Dennis G. Collins, p.4

2007 July 9. Measuring Continuous Symmetry and Positional Entropy, Poster presented at MAXENT2007, 27th Int. Workshop on Bayesian Inference and Maximum Entropy, Saratoga Springs, NY

2007 Feb. 23. An Algorithm to Measure the Symmetry of n-Points, presented at SIDIM-XXII Conference, UPR-Ponce, PR.

2007 Jan. 6. An Algorithm to Measure the Symmetry of n-Points, presented at Annual Meeting of the American Math Society, New Orleans, LA. AMS Abstract #1023-52-1900.

2006 Feb. 23. Versions of the Probability Centrifuge Algorithm, offered at SIDIM-XXI Conference, Turabo, PR.

2005 Aug. 8. Algorithm for Minimum Laterallty-Adiabatically Reduced Fisher Information, Poster presented at 25th Int. Workshop on Bayesian Inference and Maximum Entropy, San Jose State Univ., San Jose, CA.

2005 Feb. 26. Algorithm for Minimumn Laterally-Adiabatically Reduced Fisher Information, presented at SIDIM-XX Conference, UPR-Mayaguez, Mayaguez, PR.

2004 Feb. 28. From Energy to Entropy to Emergy, presented at SIDIM-XIX Conf., Cayey, PR.

2004 Jan 29. Moral Codes II, presented at 3rd Biennial Emergy Research Conference, Univ.of Florida, Gainesville, FL

2003 Nov. 14-15. Moral Codes II, presented as poster at Sigma Xi (Scientific Research Society) Annual Convention, Los Angeles, CA.

2003 Aug. 3. Divvy Economies Based on (an Abstract) Temperature, presented at MAXENT2003, 21st Int. Workshop on Bayesian Inference and Maximum Entropy, Jackson Hole, Wyoming.

2003 Feb. 21. Derivation of Vector Operators in Orthogonal Curvilinear Coordinates by Differential Forms, presented at SIDIM-18 Conf., Catholica Univ., Ponce, PR.

2002 July 11. M-Matrices and Emergy, presented at 50th Anniversary and 2002 Annual Meeting of SIAM (Soc. of Industrial and Applied Math.), Phgiladelphia, PA.

2002 Feb. 23. M-Matrices and Emergy, presented at SIDIM-17 Conference, San German, PR.

Curriculum Vitae for Dennis G. Collins, p.5

2001 Nov. 10. Thermodynamic Modeling for Social Science Problems, poster at Annual Sigma Xi (Scientific Research Soc.) Convention, Raleigh, NC

2001 Sept. 21. On the Rationale of the Transformity Method, presented at 2nd Biennial Emergy Research Conference, Univ. of Florida, Gainesville, FL.

2001 Feb. 24. Thermodynamic Modelling of Moral Codes (=Moral Codes I), presented at SIDIM-16 Conference, Humacao, PR.

2000 June 5. How to Make an Energy-Flow Digraph into a Dynamical System, presented at 9th Quadrennial Conference on Graphs, Kalamazoo, MI.

2000 Feb. 26. On the Rationale of the Transformity Method, presented at SIDIM-15 Conference, Mayaguez, PR.

1999 Sept. 4. (with Prof. H.T.Odum) Evaluating Emergy and Transformity from Energy Transformation Equations, presented at First Biennial Emergy Analysis Research Conference, Univ. of Florida, Gainesville, FL. (published 2000. Please See above.)

1999 Aug. 4. Temperature-Based Ascendancy Derived from a Cost or Reward Function, presented at MAXENT1999, 19th conference on maximum entropy and Bayesian Methods, Boise State Univ., Boise, Idaho. (published 2001. Please see above.)

1998 June 19-20. Up Multiplication Mountain, poster presented at 10th International *Mathematica* Conference, Chicago, IL.

1998 May 20. Integrative Paradigms for the Environment, presented at 3rd Regional Encounter on Education and Thought, Mayaguez, PR

1998 Feb. 28. An Approximate Least-Squares Method in a Projective-Type Space, presented at SIDIM-13 Conference, Humacao, PR.

1997 July 2. Business-Generalized Thermodynamics, presented at symposium of H.T.Odum, Univ. of Florida, Gainesville, FL, paper published later (Please see above.)

1997 May 3. On the Generalized Thermodynamics of Mime-Matter Interactions, presented at Detroit, MI meeting of the American Math Society (AMS), AMS Abstract #922-90-248.

1996 June 7. On the Thermodynamics of Mime-Matter Interactions, presented at 8th Quadrennial Int. Conf. on Graphs, Kalamazoo, MI.

Curriculum Vitae for Dennis G. Collins, p.6

1996 April 29. On the Thermodynamics of Mime-Matter Interactions, presented at SIDIM-11 Conference, Arecibo, PR

1995 Feb. 25. An Optical Echo Theory of Quasars, presented at SIDIM-10 Conf., Rio Piedras (UPR), PR.

1995 Jan. 10. An Optical Echo Theory of Quasars, presented at annual American Math Society meeting, Orlando, FL.

1994 June 23. On the Dangers of Downsizing, presented at Int. Symposium on Economic Modelling, World Bank, Washington, D.C.

1994 Feb. 26. On the Expansion Rate of the Universe, presented at SIDIM-IX Conference, Humacao, PR.

1993 June 2. Temperature for Economic Systems, presented at Conference on Energy in the Environment by H.T.Odum, Salisbury State Univ., Salisbury, MD.

1993 Feb. 7. On the Analog of Least Squares in Linear Programming, presented at SIDIM-8 Conference, San German (Interamerican Univ.), PR.

1992 Feb. 29. Entropy of Centered Gaussian Products, presented at SIDIM-7 Conference, Mayaguez, PR.

1990 June 15. Applications of a Runge-Kutta Program, presented at IMACS 1st Int. Conf. on Computational Physics, Boulder, CO.

1990 Feb. 24. A Case Study of Puerto Rican Wage-Level vs. Unemployment by Thermodynamic Modelling 1980-1988, for SIDIM-V Conf., Arecibo, PR.

1990 Jan. 19. Toward a Dispersive Number Line, presented at annual meeting of the American Math Society (AMS), Louisville, KY. AMS Abstract #854-60-321.

1987 June 19. A Generalized Inverse Example, presented at Math Assoc. of America (MAA) Conference on Linear Algebra and Graph Theory, Duluth, MN.

1987 Jan. 21. Liquid-Vapor Modeling of Merged Components for Social Science Problems, presented at annual meeting of American Math Society (AMS), San Antonio, TX. AMS Abstract #831-90-331.

1986 Jan. 9. Liquid-Vapor Models for Social Science Problems (8 Case Studies), presented at annual meeting of AMS, New Orleans, LA. AMS Abstract # 825-90-327.

Curriculum Vitae for Dennis G. Collins, p.7

1985 Jan. 9. Oil—Merger Vs. War in Iran and Iraq, presented at annual meeting of AMS, Anaheim, CA. AMS Abstract #816-90-5

1984 June. Interpolation of Repeated Exponentiation, presented at AMS meeting, Plymouth, NH. AMS Abstract #812-26-02.

1983 Dec. 5. Evolution of Computers (application of Arthur M. Young's theory), presented at INFOR-25, Mayaguez, PR.

1983 Aug. Liquid-Vapor Models (study of Braniff Airlines), paper accepted for 4th Int. Conf. on Math Modeling, Zurich, Switzerland; did not attend because of lack of support.

1982 April 29. 'Covariant' and 'Contravariant' Linear Transformations, presented at Conference on Applied Linear Algebra (SIAM), Raleigh, NC.

1982 April 16. Rules of Algebra Applied to Exponentiation, presented at AMS meeting, Madison, WS. AMS #794-26-114.

1981 July. On Some Analogies of Thermodynamics with Problems in Social Science, presented by mail/tape recorder at 3rd Int. Conf. on Math Modeling, Los Angeles, CA; did not attend personally because of lack of support.

1980 Oct. The Hudson-Nash Merger: An application of Thermodynamics to a Social Science Problem, presented at MAA (Math Assoc. of America) Indiana section meeting, Greencastle, IN.

1980 April. Contributions of Historical Examples, talk published in Symposium on the Use of History in the Teaching of Math in Memory of Arthur Hallerberg, Valparaiso Univ., Valparaiso, IN; also transcribed two of invited talks from tapes For this publication.

1978 July. A Thermodynamic Goal Model, presented at Symposium in Honor of R.J. Duffin, Pittsburgh, PA.

1975 May. Analysis of Preference Matrices (study of applied logic), Ph.D. Dissertation; IIT advisor Prof. William Darsow; evaluated by Prof. C.L. Liu, Dept. of Computer Science, Univ. of Illinois, Urbana-Champaign, IL.

1970 May. Aggregate Logic, M.S. Thesis; IIT advisor Dr. H. Ian Whitlock.

Curriculum Vitae for Dennis G. Collins, p.8

ADDITIONAL PROFESSIONAL ACTIVITIES

Memberships: 1967- . American Math Society (AMS).
                1974- . Society of Photo-optical Instrumentation Engineers (SPIE—Int. Soc. for Optical Engineering).
                1987- . Math Association of America (MAA).
                1987-1996. Asociacion Puertoriquena de Matematicas (APMM).
                1990- . Sigma Xi, Scientific Research Society.
                1993- . SIAM, Society of Industrial and Applied Mathematicians.
                1994- . NYAS, New York Academy of Sciences.
                2000- . AAAS, American Assoc. for the Advancement of Science
                2002- . INFORMS.

Research Grants: NSF Seed Money Grants of approx. $312 in 1985, $350 in 1884, And $200 in 1983 (for computer equipment).

Committees: Applied Math. Committee, 2008- .
1997-2003. Math Dept. Representative to Dialog Committee to Rector.
1999-2001. SIDIM-XV Publication Committee. Only SIDIM Conf. published so far.
1994-1996. Chairman (and Faculty Representative for for Spring Semester 1994), Math Dept. Personnel Committee.
1992. Appointed Chairman of M.S. Committee for Martin Sanchez.
1991-1992. Member of Committee to evaluate special math courses for math majors.
1986-1988. Member of Graduate Committee of Math Dept., Unov. of Puerto Rico, Mayaguez; principal architect of two M.S. Combinatorics Exams, 1988.
1983-1988 and 1995-1999. Chairman of Property Committee, Math Dept.
1988. Member of M.S. Committee for A. Gonzalez Rios, La Proyeccion Mercator in *PASCAL*.
1988 March. Faculty Advisor for Junior Technical Meeting talk by D. Feliciano, Las Formas Canonicas, Humacao, PR.

Seminars (presented): 2007 April 13. Demonstration of VQM (Visual Quamtum Mechanics) and AVQM (Advanced Visual Quantum Mechanics) by B. Thaller, Quantum Mechanics Seminar.
2002 April 18. Eigenvalue Ranking Methods, Math Dept. Colloquium, Univ. of Puerto Rico, Mayaguez, PR.
2000 Nov.16. Thermodynamic Modelling of Moral Codes (Moral Codes I), Math Dept. Colloquium, Univ. of Puerto Rico, Mayaguez, PR.
2000 Oct. 13. Quantum Probability and Computing, Parallel Algorithms and Visualization Seminar (Bollman), Math Dept., Univ. of Puerto Rico, Mayaguez., PR.

Curriculum Vitae for Dennis G. Collins, p.9

1999 Oct. 28. Temperature-Based Ascendancy and Versions of a Fourth Law of Thermodynamics, Math Colloquium, Math Dept., Univ. of Puerto Rico, Mayaguez, PR.
1998 March 12. An Approximate Least-Squares Method in a Projective-Type Space, Math Dept. Colloquium, Univ. of Puerto Rico, Mayaguez, PR 00681.
1996 July 10. Thermodynamics of Mime-Matter Interactions, at Prof. Odum's Seminar, Univ. of Florida, Gainesville, FL.
1990 March 8. A Thermodynamic View of Wage-Level Vs. Unemployment, Math Dept. Colloquium, Univ. of Puerto Rico, Mayaguez, PR.
1987 Four Lectures on the Fast Fourier Transform, Applied Math Seminar, Math Dept., Univ. of Puerto Rico, Mayaguez, PR.
1987 Sept. 24. Beyond Linear Programming, Math Dept. Colloquium, Univ. of Puerto Rico, Mayaguez, PR.
1982 May. Newton's *Principia*, History of Math Seminar, Math Dept., Valparaiso Univ., Valparaiso, IN.

1977-1979. Helped organize Applied Math Seminar for four semesters, including presentation of several talks, e.g. Connection Forms for Non-Coordinate Bases, April 23, 1979, Math Dept., Univ. of New Orleans (UNO), LA.

Continued Study (attended): 2009 May 26-29. 4th Symposium on Analysis and PDE's, Purdue Univ., West Lafayette, IN.

2007 May 27-30. 3rd Symposium on Analysis and PDEs, Dept. of Math, Purdue Univ., West Lafayette, IN.

2006 April 10-12. IBM Workshop on db2, Math Dept., UPR-Mayaguez, PR.

2005 Dec. 6-9. IBM Workshop on Rational Software Architecture, Math Dept., UPR-Mayaguez, PR.

2004 May. 2nd Symposium on Analysis and PDEs, Dept. of Math, Purdue Univ., West Lafayette, IN.

2003 July 14-25. Summer School on Applications of Advanced Mathematical and Computational Methods to Atmospheric and Oceanic Problems (MCAO-03) at NCAR (National Center for Atmospheric Research), Boulder, CO.

2003 June 1-2. Assoc. of Symbolic Logic Annual Convention, Chicago, IL.

2000 July 16-19 and 23-26. IAS/PCMI Summer School on Computational Complexity. Undergraduate Research Program. Institute for Advanced Study, Princeton, NJ.

2000 Feb. 8-9. SBIR/STTR Workshop (Small Business Innovation Research and and Small Business Technological Transfer), UPR-Mayaguez, PR.

1999 Feb. 18-19. Fundamentals of Sponsored Project Administration by The College Fund/UNCF and DoD (Dept. of Defense), Rio Piedras, PR.

1997 May 8-10. Conf. on Undergraduate Math and Science Education, Michigan State Univ., E. Lansing, MI.

Curriculum Vitae for Dennis G. Collins, p.10

1994 April 18-19. IEEE Computing Curriculum '91
The Science and Engineering Reality in '94.

1991 May 16-17. Workshop on Preparation of Research Proposals, Mayaguez, PR.

1990 Jan. MAA (Math Assoc. of America) Minicourse #2: Finite Pak (all-integer linear programming) by Bittinger and Crown, Louisville, KY.

1989 June 6-12. MAA Conf. on Expert Choice by T. Saaty, Salisbury State Univ., MD.

1987 Jan. MAA Minicourse #2: Introduction to Computer Graphics by Joan Wyzkoski, San Antonio, TX.

1986 June 1-6. NSF (National Science Foundation) Conf. on Energy in the Environment, W. Virginia Univ., Morgantown, West Virginia.

1986 Jan. MAA Minicourse #10: The Use of Computing in the Teaching of Linear Algebra by Eugene Herman, New Orleans, LA.

1984 and 1985 May. One-week seminars on INTEL chips.

1984 Artificial Intelligence and Expert Systems, 6 hrs. credit with grade A, Madrid Polytech (in Puerto Rico) by J. Pasos Sierra.

1982 July 6-16. Attended AMS-SIAM Summer Seminar on Applications of Group Theory in Physics and Math Physics, Univ. of Chicago, IL.

1972 Passed General Math Preliminary Actuarial Exam with Grade 10.

Community Service: 2005 U.S. Patent #6,970,110 Probability Centrifuge Algorithm.

2003-2004 President and Delegate to Annual Meeting, local chapter #511, Sigma Xi and Science Fair Judge at Annual Meeting Nov. 13-16, Los Angeles, CA.

2003 March. U.S. Patent #6,533,450 Bottle-type Timer with Glenn H. Collins.

2001 Created set of 40 post cards, Some Mathematicians, Physicists and Computer Developers (total 124 cards).

2000-2002 and 2004- . Treasurer, local chapter #511, Sigma Xi (Scientific Research Society).

1999 Created set of 30 post cards, Men of Math and Computer Science.

2009 Feb. 7. Judge in Computer Science and Mathematics, SESO High School, PR.

2007 Feb. 3.

2006 Feb. 4.

2005 Feb. 5.

2004 Jan. 31.

2003 Jan 25.

2002 Feb. 2.

2001 Feb. 3.

2000 Feb. 5.

1999 Jan. 30. Judge in Computer Science and Mathematics, SESO (High School) Science Fairs, Mayaguez, or Mani, PR

1987 May 13. Judge in Computer Science, 38th Int. Science & Engineering Fair, San Juan, PR. Also Judge in State Science and Engineering Fair, San German, PR.

Curriculum Vitae for Dennis G. Collins, p.11

1983 Created set of 54 post cards of mathematicians and physicists.

1979 Talk on metric system to Kiwanis Club, Chalmette, LA.

1972 Spring Quarter. Jury Duty, Lake County, IN.

Honors: Listed in 2006-2007 Cambridge Who's Who, Honors Edition, 498 RexCorp Plaza, West Tower, Uniondale, NY 11556

Marquis Who's Who in the World, 1993- .

in America, 1995- .

in Science and Engineering, 1992- .

Outstanding Man of the 21st Century (American Biographical Inst.-ABI).

2000 Outstanding Scientists of the 20th Century (Cambridge Biographical Inst.-CBI).

Who's Who of Midwest 1992.

Man of the Year 1998 (ABI).

Int. Man of the Year 1993-1994 for Services to Mathematics (CBI).

Who's Who of Contemporary Achievement 1984.

Dictionary of Int. Biography 1980.

Who's Who of South & Southwest 1978-1979 and 1997-1998.

SOME CONTINUOUS EMPOWER2 MODELS

By Dennis G. Collins
Dept. of Math Sciences, UPR-Mayaguez
Box 9018
Mayaguez, PR 00681-9018
Sept. 12, 2009

ABSTRACT: There is much interest in how various species compete, as well as, for example, energy start-up companies in business. This study covers five levels of competition for a resource base, derived mathematically from the continuous logistic or S-curve model: 1) logistic curve with powers, 2) competition with different powers, 3) competition with "triple" or "quadruple" helix effect, and 4) competition with also energy losses and 5) powers versus helix versus both. It is found that the most successful competitor is highly dependent on initial conditions versus the "tipping point." Also the "triple helix" can propel competitors with higher powers to the forefront. Putting in energy losses as required by the 2nd law of thermodynamics, curiously even further advances higher-power competitors. Having both powers and helix is best. It remains to bring in consideration of pulsing and transformity.

INTRODUCTION

Typically empower is considered in a thermodynamic setting, as emergy per unit time. However H.T. Odum considered a second meaning in his Salisbury State University lectures in June, 1993, namely empower could be related to the mathematical power of feedback interaction, i.e. be related to the mathematical term "nth power" as a product of n factors. He seemed to view this possibility as a hidden and yet-unexplained driver of the "maximum empower" principle. In semantic notation this meaning could be denoted empower with a subscript 2, and it is the topic of this study.

H.T. Odum's discrete models (Odum 1983) can cover all cases; however the study of continuous models with differential equations seems to give some advantages of analysis. Here five levels of continuous models are derived from the logistic curve (Andrews, 1991) or S-curve: 1) feedback powers can be put into the logistic curve, 2) a system of differential equations can be set up to see how the different powers compete, 3) the powers can change to a "triple or quadruple helix" feedback, and 4) the system can be equipped with energy losses, as with H.T. Odum's models. Finally 5) both powers and helix can be compared against each individually. The five cases are considered in the next five sections. Mathematical powers are indicated by the "^" symbol, for example x*x*x=x^3.

DIFFERENT FEEDBACK POWERS

The one-dimensional logistic or S-curve, with limit to growth "a," has the simplified equation dy/dt= y` = y (a-y). The equation can be solved with higher power of "n."

Differential equation	Solution (t in terms of y)
y`= y (a-y)	(1/a) (*ln* (y/(a-y)))
y`= y^2 (a-y)	(1/a^2) (*ln* (y/(a-y)) -a/y)
y` = y^3 (a-y)	(1/a^3) (*ln* (y/(a-y)) -a/y - a^2/(2 y^2))

If the expressions for y are graphed versus t or time, they all have the form of S-curves, with, however, steeper slopes the larger the power of y on the right-hand side of the differential equation. This increasing slope means that the competitor with higher power should be able to reach the resource limit "a" faster. Please see Program 1 and Graph 1.

COMPETITION AMONG DIFFERENT POWERS

To see how competitors with different feedback powers compete, a system of differential equations can be set up as follows:

Y1' = Y1 (a - (Y1+ Y2 +Y3))
Y2' = Y2^2 (a - (Y1+ Y2 +Y3))
Y3' = Y3^3 (a - (Y1+ Y2 +Y3)).

In this case the three competitors compete for a resource base "a". The selection of the values for "a" and the system's initial conditions turn out to be very important. Here "a" is selected to be 10, because 1 is a mathematical "tipping point" (Gladwell 2000) for powers, and it is desirable to consider changes of y across 1, say from 0 to 10. Powers y^n below 1 get smaller as n increases, but powers above 1 get larger. For example (.5)^3 = 0.125 < 0 .5 < 1 but 2^3 = 8 > 2 > 1.

The system above with a=10 can easily be solved numerically by the *Mathematica* programming language with various initial conditions. It is found that with initial conditions Y1(0) =.0.2, Y1(0)=0.2, Y3(0) =0.2, the supposedly higher-power competitors never reach 1 and Y1 controls more than 8 of the 10 resource units as time increases. Please see Program 2 and Graph 2. On the other hand if the initial conditions are set at Y1(0) =2, Y2(0) =2, Y3(0) = 2, above the tipping point of y = 1, then Y3 quickly reaches about 5.3 resource units, Y2 controls about 2.7 units and Y1 controls about 2.5 units. Please see Program 3 and Graph 3. Most of the advantage in gaining resures or "high-power talent" will be suppressed unless the higher-power component somehow obtains more than 1 unit initially. So far, the author has not found equations as above in any differential equations book, although the powers are related to what is called the order of chemical reactions. The same result applies to chemistry books. Logofet (1993) covers mostly quadratic interactions. Remark: The elusive "tipping point" can depend in general on many factors, including the competition.

COMPETITION WITH "QUADRUPLE HELIX" (HOME RUN) FEEDBACK

The term "triple helix" is applied here to mean the combined effect of many feedback streams (Lewontin 2000 and Collins and Scienceman 2008). If the input streams are above the "tipping point" of 1, there can be a high-power effect as in y^n. "Triple-helix" feedback would replace $Y2^2$ in the above equations with $Y1*Y2$ and $Y3^3$ by $Y1*Y2*Y3$. However the results are easier to track with four competitors, so that the model is extended to a four-dimensional model without much loss, as follows:

$Y1' = Y1$ $(10 - (Y1 + Y2 + Y3 + Y4))$

$Y2' = Y1*Y2$ $(10 - (Y1 + Y2 + Y3 + Y4))$

$Y3' = Y1*Y2*Y3$ $(10 - (Y1 + Y2 + Y3 + Y4))$

$Y4' = Y1*Y2*Y3*Y4$ $(10 - (Y1 + Y2 + Y3 + Y4))$

Here instead of "higher-power feedback" there is feedback from all previous units. It is questionable if Y4 has more talent or merely more acclaim by reason of input from other competitors.

The results show very-sensitive dependence on initial conditions. For example with initial conditions $Y1(0) = 0.1$, $Y2(0) = 0.1$, $Y3(0) = 0.1$, $Y4(0) = 0.1$, the "quadruple-helix candidate" never makes it above the tipping point 1, and the 10 resource units are divided roughly as $Y1 = 3.6$, $Y2 = 3.3$, $Y3 = 2.2$ and $Y4 = .9$ as time increases. Please see Program 4 and Graph 4.

On the other hand with initial conditions set at $Y1(0) = 0.4$, $Y2(0) = 0.4$, $Y3(0) = 0.4$, $Y4(0) = 0.4$, there is a crossover point at which all four competitors control about the same amount of the resource (about 2 units) and beyond which the competitors reverse order to reach limits of $Y1 = 2.1$, $Y2 = 2.2$, $Y3 = 2.5$, $Y4 = 3.2$. Please see Program 5 and Graph 5. Thus, even from below 1 initial conditions, "quadruple-helix" can get a boost from the other candidates over the "tipping point" of 1, and go from there to exert maximal resource control. In baseball, a player who hits a home run benefits from all other teammates on base at the time.

ABOVE COMPETITION WITH ENERGY SINKS

H.T. Odum always insisted on putting energy sinks on models to reflect the operation of the 2nd law of thermodynamics that affects all processes, so that it is desirable to consider what happens if energy sinks are added to the above model as follows :

$Y1' = Y1$ $(10 - (Y1 + Y2 + Y3 + Y4)) - .1\,Y1$

$Y2' = Y1*Y2$ $(10 - (Y1 + Y2 + Y3 + Y4)) - .1\,Y2$

$Y3' = Y1*Y2*Y3$ $(10 - (Y1 + Y2 + Y3 + Y4)) - .1\,Y3$

$Y4' = Y1*Y2*Y3*Y4$ $(10 - (Y1 + Y2 + Y3 + Y4)) - .1\,Y4$

Please see Program 6 and Graph 6. The results seems to indicate not much change over the beginning of the evolution, or much dependence on the energy sink coefficient, set here equal to 0.1, but after the 10 units are distributed there is a tendency of the maximal competitor to drive the others down to the "tipping point" of 1. In the last mentioned model above, Y4 goes from about 3.2 to 7, and Y1, Y2 and Y3 down to about 1, as time increases.

Please see Program 7 and Graph 7. Thus the "quadruple-helix" candidate extends his control. In fact the "quadruple-helix" factors have more or less vanished into factors of 1, except for Y4. To respond to a reviewer comment, this trend seems to carry over into spatial extent as expressed by the following statement (Wayne 2010):

". . . even within the long-term regime, there are two distinct time-scales, one on which the inviscid phenomena predicted by Onsager appear and a second, typically longer, time scale over which viscous effects manifest themselves and which we show below lead to the formation of a single vortex in the system." Here "inviscid" means no second-law energy sinks as in the third case above. There seems to be a good chance of extending spatially the competition discussed here by working with vortices in 2-dimensional space.

COMPETITION OF POWERS VERSUS HELIX VERSUS BOTH EFFECTS

Please see Program 8 and Graph 8. As the reader may suspect, having both effects will allow a competitor, say Y4 in the following system, to dominate long:

$$Y1' = Y1 \qquad (10 - (Y1 + Y2 + Y3 + Y4)) - .1\ Y1$$
$$Y2' = Y1*Y2 \qquad (10 - (Y1 + Y2 + Y3 + Y4)) - .1\ Y2$$
$$Y3' = Y3^2 \qquad (10 - (Y1 + Y2 + Y3 + Y4)) - .1\ Y3$$
$$Y4' = Y1*Y2*Y4^2 \qquad (10 - (Y1 + Y2 + Y3 + Y4)) - .1\ Y4$$

With initial conditions $Y1(0)=0.1$, $Y2(0)=0.1$, $Y3(0)=0.1$, $Y4(0)=0.1$, the limits show Y4 controlling about 8.5 units of 10, with Y1 left with about 1 unit and Y2 about 0.4 and Y3 about 0.1. However Y2 does get above 6 and leads until about 10 units. Y3 gets to about 1 before fading. Please see Program 9 and Graph 9. However under unequal initial conditions, with say $Y3(0)$ (only) changed to 2 units, above the tipping point, Y3 quickly dominates the entire 10 resource units.

CONCLUSIONS

The above study leaves many questions open, such as, Where is the "tipping point?" and "Does it exist in the real world? Program 10 and Graph 10 illustrate the effect of not including competitor 4 in the limits on resources. There is also the question of how to extend the model to pulsing, since the models studied are like trial heats at a track meet, which end up with a limiting value for each competitor. Please see Program 11 and Graph 11 for a (more-or-less standard) 2-competitor pulsing model. However the models may form some basis for the study of feedback from intrinsic talent versus feedback from other competitors. It seems intrinsic feedback power Y^n may get replaced by the "multiple helix effect" of $Y1*Y2*\ldots$ Yn over time by a process of competition, whereby some competitors may lose out factors to others they consider more talented, and some with multiple-feedback talent may "rest on their laurels," replacing self feedback with feedback from others. There is also a Biblical lesson that the most talented may be suppressed without a spirit of cooperation. A reviewer says no actor in the model shows a spirit of cooperation. However the fact that, say, the fourth equation has the factor $Y1*Y2*Y3*Y4$ means that Y4 is depending on Y1 to Y3. If any of these other factors Y1 or Y2 or Y3 go to zero, so does Y4. Presumably Y1 or Y2 or Y3 could eventually prevent their factors from being included in the Y4 equation. This dependence is a main meaning of the "triple helix" effect. Presumably negative terms including products of factors could also be put into the equation expressing destructive interaction instead of cooperation as is done routinely in ecological models. It may be hoped the models somehow help to sort out the various proposed solutions or create new solutions to the energy and other problems facing society. Of course the models can be modified in many ways. The pure y^n talent is the more likely to be suppressed the larger the value of n, unless artificially started above the "tipping point," although the whole theory may be said to be related to whether the factors multiplying y (namely the remaining y factors and the remaining part of the resource "pie") come out with product bigger than that of competitors versus their variable. A reviewer mentions that study of the tipping point may be helped by considering fractional powers in the equations. This procedure leads to interpolated results. For example solving the differential equation $y` = y^{(3/2)}* (a-y)$ yields a *Mathematica* solution $t= 2*ArcTanh[Sqrt[y/a]]/a^{(3/2)}-2/(a*Sqrt[y])$, which with a=10 and graphed as y versus t crosses the y-axis about y(0)=7 between the value y(0)= 5 for the logistic curve (n=1) and the value about y(0)=8 for the quadratic (n=2) power of y in the equation $y` = y^2 (a-y)$.

ACKNOWLEDGMENTS

The author acknowledges many discussions with H.T. Odum, as well as a Sigma Xi Distinguished Lecture on "Tipping points in academe" by Daryl Chubin, from AAAS May 14, 2009 at Mayaguez, PR, sponsored by the Chancellor's Office at University of Puerto Rico, Mayaguez. Also many thanks are due reviewers. A caution is that the results are dependent on the *Mathematica* programs working correctly.

REFERENCES

Andrews, Larry 1991. *Introduction to Differential Equations*, Harper-Collins, NY. poster at 5th Biennial Emergy Conference, and submitted for inclusion in Proceedings.

Collins, Dennis and Scienceman, David, Political spectrum models. In: Brown, M.T. (Ed.) *Emergy Synthesis* 5: Theory and Applications of the Emergy Methodology. The Center for Environmental Policy, Gainesville, pp.23-33.

Gladwell, Malcolm 2000. *The Tipping Point*, Little, Brown and Company, New York.

Lewontin, Richard 2000. *The Triple Helix*, Harvard University Press, Cambridge, MA.

Logofet, Dimitrii 1993. *Matrices and Graphs*, CRC Press, Boca Raton, FL.

Odum, H.T. 1983. *Systems Ecology*, John Wiley and Sons, New York.

Wayne, C. Eugene 2010. Vortices and Two-Dimensional Fluid Motion, *Notices of the American Mathematical Society*, Vol. 58 No.1, p.13.

APPENDIX

```
Clear All
A=10;
T1=Table[{(1/a)*Log[y/(a-y),y},{y,.1,9.9,.1}];
T2=Table[{(1/a^2)*(Log[y/(a-y)]-a/y},{y,.1,9.9,.1}];
T3=Table[{(1/a^3)*(Log[y/(a-y)]-a/y-a^2/(2*y^2)),y},{y,.1,9.9,.1})];

ListPlot[T1]
ListPlot[T2]
ListPlot[T3]
Program 1. Formulas for power logistic growth.
```

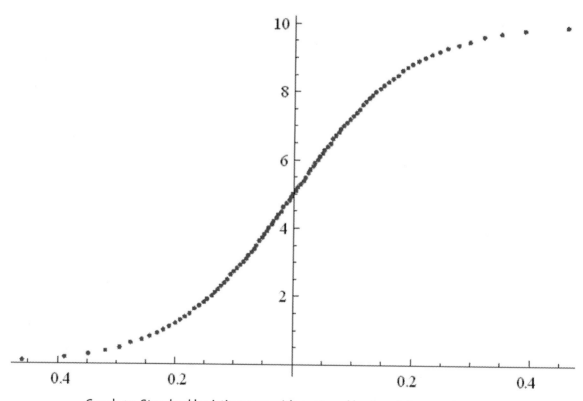

Graph 1a. Standard logistic curve with a=10 and horizontal asymptote y=0.

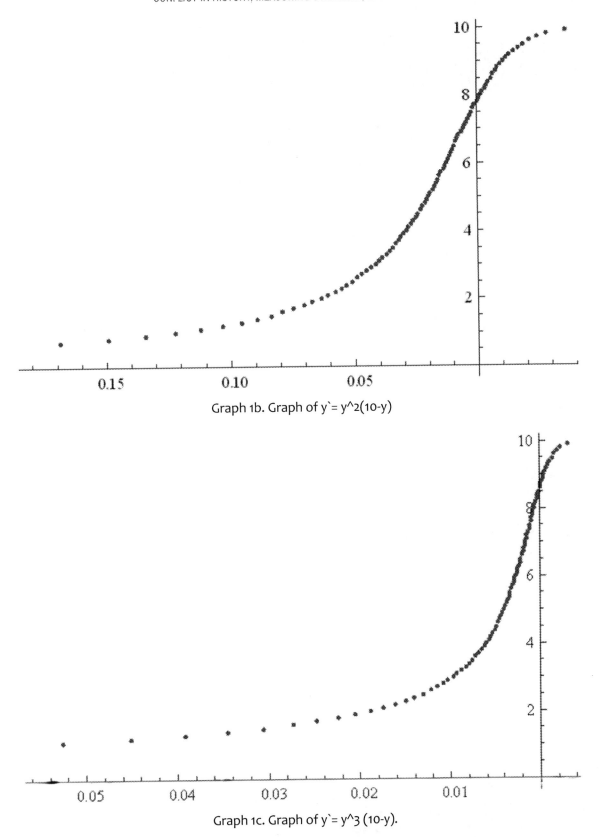

Graph 1b. Graph of y`= y^2(10-y)

Graph 1c. Graph of y`= y^3 (10-y).

Graph 1. The standard logistic curve (1ˢᵗ power, top Graph), quadratic power growth (middle), and cubic growth (bottom).

Clear All
s=NDSolve[{y1'[t]y1[t]*(10.0-(y1[t]+y2[t]+y3[t])),y2'[t](y2[t])^2*(10-
(y1[t]+y2[t]+y3[t])),y3'[t](y3[t])^3*(10-(y1[t]+y2[t]+y3[t])),y1[0].2,y2[0].2,y3[0].2},{y1,y2,y3}
,{t,20}]

Program 2. Competition among different powers.

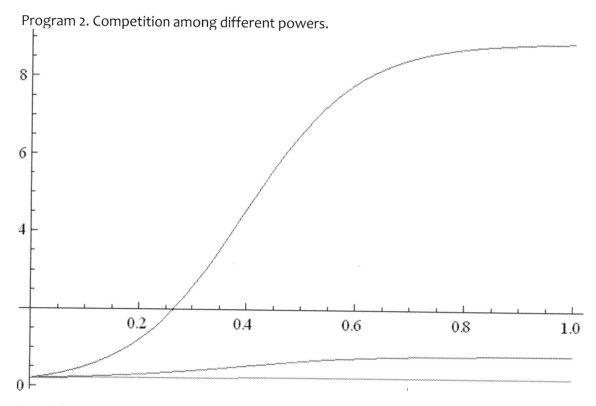

Graph 2. Competiton among different powers, initial conditions (=.2) below tipping point. 1st power top graph, 2nd power middle, 3rd power (cubic) bottom.

Clear All
s=NDSolve[{y1'[t]y1[t]*(10.0-(y1[t]+y2[t]+y3[t])),y2'[t](y2[t])^2*(10-
(y1[t]+y2[t]+y3[t])),y3'[t](y3[t])^3*(10-(y1[t]+y2[t]+y3[t])),y1[0]2,y2[0]2,y3[0]2},{y1,y2,y3},{
t,20}]

Program 3. Competition among powers, initial conditions above tipping point.

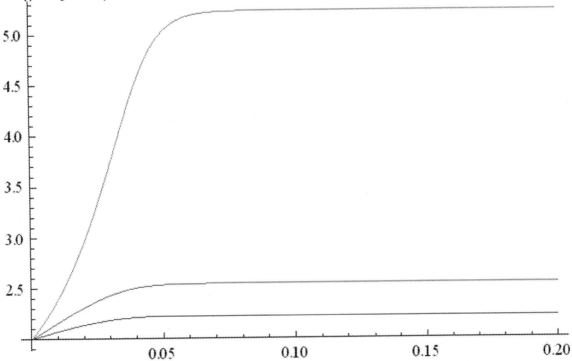

Graph 3. Competition among powers, initial conditions (=2) above tipping point.
Top graph is now 3rd power (cubic), middle graph is 2nd power (quadratic) and bottom graph is 1st power (linear).

Clear All
s=NDSolve[{y1'[t]y1[t]*(10.0-(y1[t]+y2[t]+y3[t]+y4[t])),y2'[t](y2[t]*y1[t])*(10-
(y1[t]+y2[t]+y3[t]+y4[t])),y3'[t](y3[t]*y2[t]*y1[t])*(10-(y1[t]+y2[t]+y3[t]+y4[t])),
y4'[t](y1[t]*y2[t]*y3[t]*y4[t])*(10-(y1[t]+y2[t]+y3[t]+y4[t])),y1[0].1,y2[0].1,y3[0].1,y4[0].1},{
y1,y2,y3,y4},{t,20}]

Program 4. Competition with "quadruple helix" logistic differential equations system, initial conditions (=.1) below tipping point.

Plot[Evaluate[{y1[t],y2[t],y3[t],y4[t]}/.s],{t,0,1}, PlotRange→All]

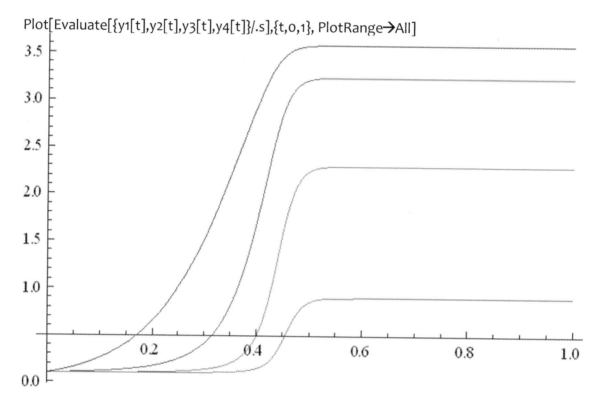

Graph 4. Competition with "quadruple helix" logistic differential equations system, initial conditions (=.1) below tipping point. Top graph is 1st power, next is product of 1st and 2nd, then product of 1st, 2nd and 3rd, finally bottom graph is product of all four.

Clear All
s=NDSolve[{y1'[t]y1[t]*(10.0-(y1[t]+y2[t]+y3[t]+y4[t])),y2'[t](y2[t]*y1[t])*(10-
(y1[t]+y2[t]+y3[t]+y4[t])),y3'[t](y3[t]*y2[t]*y1[t])*(10-(y1[t]+y2[t]+y3[t]+y4[t])),
y4'[t](y1[t]*y2[t]*y3[t]*y4[t])*(10-(y1[t]+y2[t]+y3[t]+y4[t])),y1[0].4,y2[0].4,y3[0].4,y4[0].4}
,{y1,y2,y3,y4},{t,20}]

Program 5. "Quadruple helix" with initial conditions (=.4) above tipping point. Observe tipping point is now less than 1 because of "teamwork"

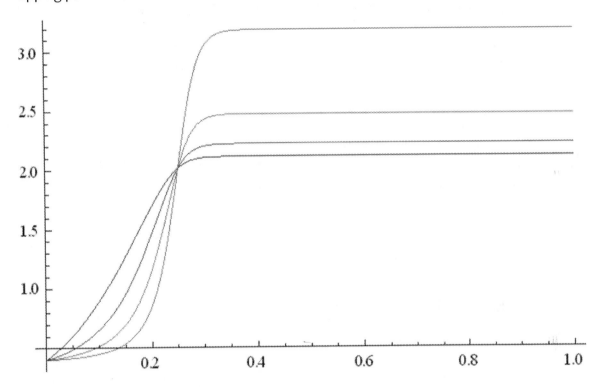

Graph 5. "Quadruple helix" with initial conditions (=.4) above tipping point. Order of competitors reverses, so that on right (at t.=1.0) 4th competitor is top, then 3rd, then 2nd, and finally 1st.

Clear All

s=NDSolve[{y1'[t]y1[t]*(10.0-(y1[t]+y2[t]+y3[t]+y4[t]))-.1*y1[t],y2'[t](y2[t]*y1[t])*(10-(y1[t]+y2[t]+y3[t]+y4[t]))-.1*y2[t],y3'[t](y3[t]*y2[t]*y1[t])*(10-(y1[t]+y2[t]+y3[t]+y4[t]))-.1*y3[t],

y4'[t](y1[t]*y2[t]*y3[t]*y4[t])*(10-(y1[t]+y2[t]+y3[t]+y4[t]))-.1*y4[t],y1[0].4,y2[0].4,y3[0].4,y4[0].4},{y1,y2,y3,y4},{t,20}]

All Clear

{{y1→InterpolatingFunction[{{0.,20.}},<>],y2→InterpolatingFunction[{{0.,20.}},<>],y3→InterpolatingFunction[{{0.,20.}},<>],y4→InterpolatingFunction[{{0.,20.}},<>]}}

Program 6. "Quadruple helix" with initial conditions (=.4) above tipping point and energy sinks (= .1*quantity), viewed over shorter time interval (0 <= t <= 1).

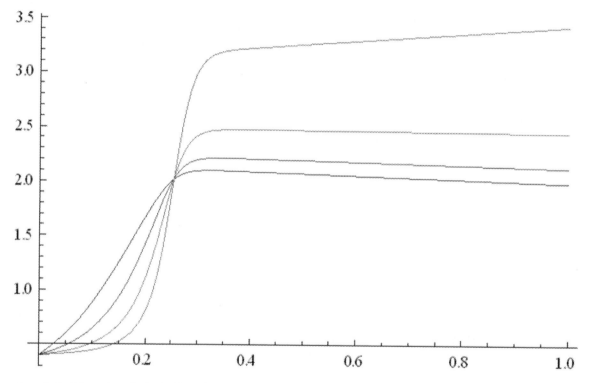

Graph 6. "Quadruple helix" with initial conditions (=.4) above tipping point and energy sinks (=.1*quantity), viewed over shorter time interval (0<=t<=1). At right (t=1) top graph is 4th competitor, next 3rd, then 2nd, and finally bottom is 1st competitor.

Clear All
s=NDSolve[{y1'[t]y1[t]*(10.0-(y1[t]+y2[t]+y3[t]+y4[t]))-.1*y1[t],y2'[t](y2[t]*y1[t])*(10-
(y1[t]+y2[t]+y3[t]+y4[t]))-.1*y2[t],y3'[t](y3[t]*y2[t]*y1[t])*(10-(y1[t]+y2[t]+y3[t]+y4[t]))-
.1*y3[t],
y4'[t](y1[t]*y2[t]*y3[t]*y4[t])*(10-(y1[t]+y2[t]+y3[t]+y4[t]))-.1*y4[t],y1[0].4,y2[0].4,y3[0].4,
y4[0].4},{y1,y2,y3,y4},{t,100}]

Program 7. Same as Program 6.

Plot[Evaluate[{y1[t],y2[t],y3[t],y4[t]}/.s],{t,0,100}, PlotRange→All]

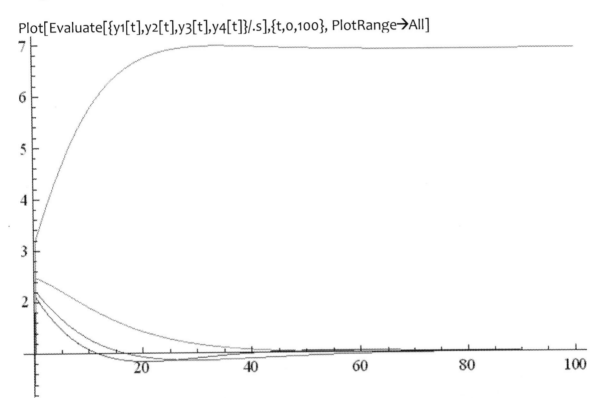

Graph 7. Results of Program 6, viewed over long time interval (0<=t<=100). 4th competitor goes from about 3.2 to 7 and other three competitors are reduced to around 1.

Clear All

s=NDSolve[{y1'[t]y1[t]*(10.0-(y1[t]+y2[t]+y3[t]+y4[t]))-.2*y1[t],y2'[t](y2[t]*y1[t])*(10-(y1[t]+y2[t]+y3[t]+y4[t]))-.2*y2[t],y3'[t](y3[t]^2)*(10-(y1[t]+y2[t]+y3[t]+y4[t]))-.2*y3[t],y4'[t](y1[t]*y2[t]*y4[t]^2)*(10-(y1[t]+y2[t]+y3[t]+y4[t]))-.2*y4[t],y1[0].4,y2[0].4,y3[0].4,y4[0].4},{y1,y2,y3,y4},{t,100}]

Program 8. Competition of powers (competitors 1 and 3) versus helix (competitor 2) versus both effects (competitor 4) with energy sinks at .2*quantity.

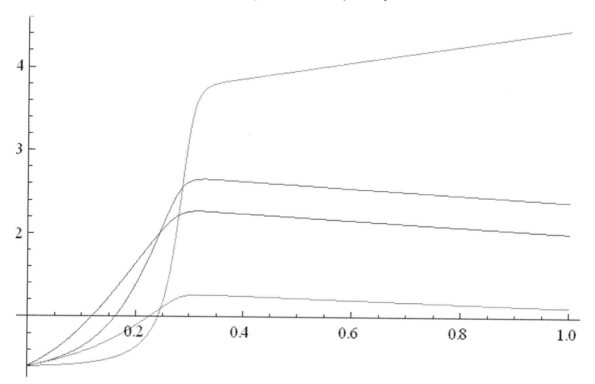

Graph 8. Competition of powers (competitors 1 and 3) versus helix (competitor 2) versus both effects (competitor 4), viewed over short time interval (0<=t<=1). By t=1.0 at right, 4th competitor is top graph, then 2rd competitor, then 1st competitor, then 3rd competitor at bottom.

Clear All
s=NDSolve[{y1'[t]y1[t]*(10.0-(y1[t]+y2[t]+y3[t]+y4[t]))-.2*y1[t],y2'[t](y2[t]*y1[t])*(10-
(y1[t]+y2[t]+y3[t]+y4[t]))-.2*y2[t],y3'[t](y3[t]^2)*(10-(y1[t]+y2[t]+y3[t]+y4[t]))-.2*y3[t],
y4'[t](y1[t]*y2[t]*y4[t]^2)*(10-(y1[t]+y2[t]+y3[t]+y4[t]))-.2*y4[t],y1[0].1,y2[0].1,y3[0].1,y4[0
].1},{y1,y2,y3,y4},{t,100}]

Program 9. Same as Program 8 except intial conditions set at .1.

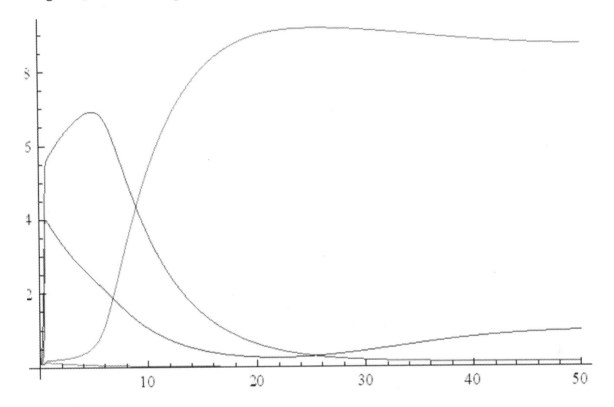

Grasph 9. Results of Program 9 viewed over long time interval (0<=t<=50). At t=50 at right, top graph is competitor 4, then competitor 1, then competitor 2, and competitor 3 (not visible along axis).

Clear All
s=NDSolve[{y1'[t]y1[t]*(10.0-(y1[t]+y2[t]+y3[t]+y4[t])),y2'[t](y2[t]*y1[t])*(10-(y1[t]+y2[t]+y3[t]+y4[t])),y3'[t](y3[t]*y2[t]*y1[t])*(10-(y1[t]+y2[t]+y3[t]+y4[t])),y4'[t](y1[t]*y2[t]*y3[t]*y4[t])*(10-(y1[t]+y2[t]+y3[t])),y1[0].2,y2[0].2,y3[0].2,y4[0].2},{y1,y2,y3,y4},{t,20}]

Program 10. "Quadruple helix" without y4 included in resource limitation on y4 equation (and no energy sinks).

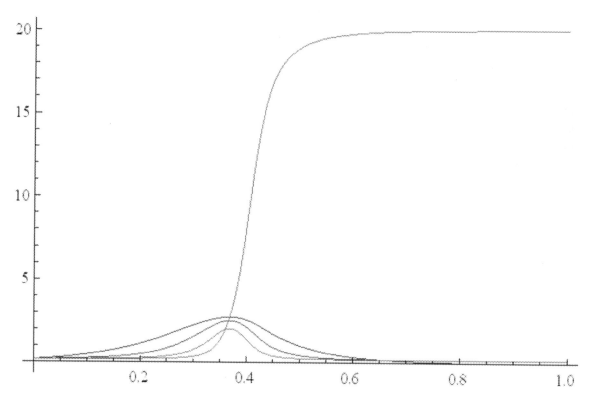

Graph 10. Results of program 10 ("quadruple helix" without energy sinks) without y4 included in limit of 10 on resources. 4th competitor reaches double resource limit and other competitors are reduced to values less than 1. Among "bell-shaped" curves, top is 1st competitor, then 2nd competitor, then 3rd competitor.

```
Clear All
s = NDSolve[{y1'[t] == y1[t]*(3 - 3*y2[t]),
y2'[t] == y2[t]*(3*y1[t] - 12)*(10 - y1[t] - 2*y2[t]), y1[0] == 8
 y2[0] == 3}, {y1, y2}, {t, 10}]

Plot[Evaluate[{y1[t],y2[t]}/.s],{t,0,2}, PlotRange→All]
```

Program 11. Example of pulsing with 2 competitors. The "2" coefficient in the second equation can be considered as a transformity.

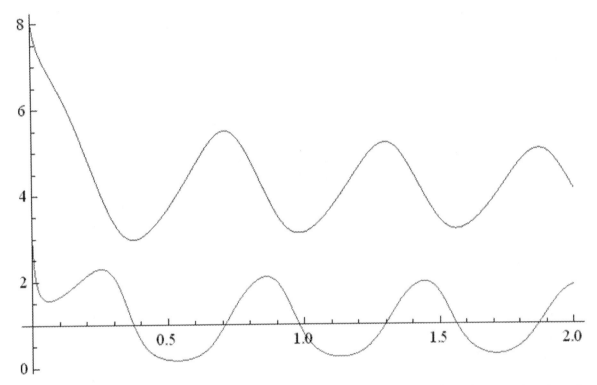

Graph 11. Pulsing with 2 competitors according to Program 11. Top graph is 1st competitor (prey).

EXAMPLES OF MEASURING CONTINUOUS SYMMETRY

By Dennis G. Collins
Dept. of Math Sciences, UPRM Box 9018, Mayaguez, PR 00681-9018

ABSTRACT: As a graduate student at Illinois Institute of Technology in Chicago, Illinois in the early 1970's, I attended a couple lectures by Karl Menger, who kindly gave me a few of his papers. He was famous for his work on continua and curves, so that I always had the goal of doing some work in this area. At first I tried to do approximations with a discrete set of points; however it was difficult to tell how good the approximations were. The present paper is based on measuring the symmetry of some elementary planar figures by doing "exact" integrals over a triangular region, which represents the set of all possible distances from any point to any other point on the figure, and measures the amount of "wiggle room" at any given distance. Calculations are carried out for straight line, square, equilateral triangle and circle, among others.

INTRODUCTION

This paper continues the process of computing continuous symmetry for various geometric figures, mostly plane figures with perimeter of length 1. There are several levels of computation at present: 1) figures for which exact values can be obtained in terms of e, π, and so on. These values are typically found by the Integrate[ ] command in *Mathematica.* 2) figures for which numerical approximations can apparently be obtained to several significant digits; typically these values are found by the NIntegrate[ ] or N[ ] command in *Mathematica.* 3) figures for which integrals can be written but do not calculate out, 4) figures for which some approximations can be found and 5) figures for which some bounds may be given . The goal is to move calculations to a smaller level number.

METHODS

There are several steps to approximating the continuous symmetry.
1) drawing the figure of perimeter 1,
2) calculating the distances on the interior and boundary of the triangle bounded by the x-axis, the y-axis and the line x+y=1. As the first parameter s is traced along the figure, its distance is measured downward from the point (0,1) to (0,0). As the second parameter t is traced along the figure, its distance is measured right from (0,0) to (1,0). As a consequence any point on the line x+y=1 represents the same distance

traced out with respect to each parameter s and t, and thus has 0 distance from the first point to the second. The distance u(x,y)= u(t,1-s) on the interior of the above triangle is calculated (with s=1-y and t=x) as the square root of (x2(t)-x1(s))^2+(y2(t)-y1(s))^2, where P(s) = (x1(s),y1(s)) and P(t) = (x2(t),y2(t)) are points on the original drawing, i.e. original coordinate system.

3) Integrating area versus u on the interior of the above triangle. As u goes from 0 to maximum distance apart of any two points on the drawing, the area goes from 0 to ½. © 2008 Dennis G. Collins

4) Multiplying the function of part 3) by 2, so that it goes from 0 to 1 as distance between points goes from 0 to maximum. This function represents the c.d.f or cumulative distribution function of a probability density.

5) Taking the derivative with respect to u of the cdf gives this density f(u).

6) Calculating the continuous entropy of f(x) or integral of - p(x) ln(p(x)) from 0 to maximum distance,

7) Taking the negative of the entropy or negentropy to get the continuous symmetry or order.

RESULTS

Results for some examples are presented as follows (exact value given after result if known):

Figure	Continuous Symmetry
1) straight line (length 1)	.193= ln(2)-1/2
2) L-shape (1/2 length each side)	.476
3) V-shape (60°, ½ length each side)	.741
4) coinciding lines (1/2 length each)	.886=ln(4)-1/2
5) square (1/4 length each side)	1.068
6) equilateral triangle (1/3 length each side)	1.161
7) circle (circumference 1, radius 1/(2*π)	1.289=2*ln(π)-1

CIRCLE THEOREM

The above results can be understood somewhat by what I call the "Circle Theorem," which is different from the "Gershgoren Circle Theorem" of eigenvalues.

This result says that if the figure is confined to within a circle of diameter d, then the order or continuous symmetry is greater than or equal to ln(1/d), due to the fact that entropy is maximized by the uniform density. The result gives a lower bound on continuous symmetry. For example, the circle of case 7) has a diameter d=1/π, leading to a continuous symmetry (1.289) greater than or equal to ln(π) = 1.114. The straight line of case 1) has diameter 1 leading to a continuous symmetry (.193) greater than or equal to ln(1) = 0. The equilateral triangle of case 6) has continuous symmetry (1.161) greater than or equal to ln(1/(1/3))= ln(3) = 1.098.

The result also carries over to three or higher dimension. For example if a three-dimensional curve is confined within a sphere of diameter d, then the three-dimensional continuous symmetry is greater than or equal to ln(1/d), since the 1-dim distance integral only goes from 0 to d and the entropy is maximized by the uniform density over this interval.

MAXIMUM ENTROPY

Surprisingly it appears that the maximum entropy or minimum continuous symmetry of a one-dimensional connected figure of length 1 must occur for a straight line, i.e. equal .193. This result happens because any "kinks" put into the straight line reduce the distance integral to a length less than one, thereby increasing the "wiggle room" of some intermediate distances. Caution: There might be a counterexample to this result. Clearly if the figure is not connected, it may extend over more distance and the entropy may go up, or the continuous symmetry go down. For example the continuous symmetry of a broken line, with each ½-length segment separated by 0.1 unit along a straight line (i.e. from 0 to .5 and then from .6 to 1.1), goes down to .0683 < .193, and if the two 1/2 length segments are separated by ½ unit along a straight line (i.e. from 0 to .5 and then from 1 to 1.5), the continuous symmetry goes down to -.1534 < .193. This value stays the same as the ½-length line segments are further separated along a straight line.

It appears if a straight line of length one is crumpled up, to straighten it out the first place to look would be the places of highest occupational numbers $f(x)$ on the above-mentioned density graph.

Similarly the continuous symmetry of a two-dimensional surface could be measured by constructing distances between pairs of points within a four-dimensional structure (simplex). Then the maximum entropy or minimum continuous symmetry of a disc (of area 1 say) would be a flat disc. If a flat disc were crumpled up, the first place to look to straighten it out would again be the places of maximum occupational number on the corresponding density.

As so often happens with symmetry definitions, it appears a crumpled connected line of length 1 will have more continuous symmetry than a straight line according to this definition.

PARALLEL LINE SEGMENTS

Another perhaps counter-intuitive result of the given defintion of continuous symmetry is that ½-length tandem parallel line segments increase in continuous symmetry as they are moved apart. This calculation happens because the dispersion of lengths decreases with increasing separation according to the Pythagorean formula (e.g. $(1^2 + 0.1^2)^{.5} = 1.00498$ but $(1000^2 + 0.1^2)^{.5} = 1000.000005$). This fact of putting more probability within a smaller interval makes the entropy go down and the continuous symmetry go up, as evidenced by the following table:

Tandem parallel lines of lenth ½	Continuous symmetry
Coinciding line segments (as above)	.886
Separated by 1 unit (say {(u,0), u e [0,.5]} and {(v,1), v e [0,.5]}	1.297
Separated by 2 units	1.638
Separated by 10 units	2.441
Separated by 100 units	3.592
Separated by 1000 units	4.743

For reference the two perpendicular ½-length line segments {(u,0),u e [0,.5]} and {(0,1+v), v e [0,.5] are calculated to have continuous symmetry .059.

CONCLUSION

As opposed to discrete symmetry as discussed previously by the author, continuous symmetry has many fascinating quirks, requiring further study. However these quirks, such as the fact that compressing a line into a smaller and smaller ball makes the continuous symmetry go to infinity, may have some application in physics and biology. Remark: By adding a ln constant term, continuous symmetry can be normalized to maximum distance 1 versus perimeter 1.

REFERENCES

2007 Colllins, Dennis G., "Algorithm to Measure Symmetry and Positional Entropy of n-Points," (talk at 2007 annual meeting of the American Mathematical Society, Jan.6, 2007 and SIDIM XXII), published in (ISSS—Int. Society of Systems Sciences) *General Systems Bulletin, Vol. XXXVI*, p.15-21.

2007 Collins, Dennis G., "Measuring Continuous Symmetry and Positional Entropy," poster presented at MAXENT2007, Saratoga Springs, NY, July 9, 2007 (unpublished manuscript).

February 29, 2008
For SIDIM XXIII, UPR-Carolina

Music of Dennis G. Collins Hilltop Records www.hilltoprecords.com
7108 Grand Blvd., Hobart, IN 46342-6628 787-951-4208

1. Music of America   HART-124   Side 1: Track 3   3:20
   Short Columbus Cantata   Charise Shelly
   (Hollywood Artists Record Company, 6000 Sunset Blvd. Ste 207, Hollywood, CA 90028)

2. America   AM-138   Side B: Track 5   2:45   1994
   Short Spaceship Cantata   Rusty Stratton
   (Hilltop Records, 6201 Sunset Blvd. #4, Hollywood, CA 90028)

3. Hilltop Country   HC-120   Side A: Track 7   2:42   1995
   Short Spaceship Cantata #3   Rusty Stratton
   (3. through 7., Hilltop Records, 1777 No. Vine St. 321, Hollywood, CA 90028)

4. Land That I Love   LL-101   Track 9   2:25   2002
   Short Cosmic Cantata   Cody Lyons

5. Sing Hosanna   XM-159   Track 23   2:38   2005
   Christmas Values   Cody Lyons
   (part of One Size Fits All)

6.	Hurricane New Orleans Serenade	HKD-102	Track 9 Cody Lyons	2:29	2006
7.	The Magic of Christmas Christmas and Amadi's Lullaby	XM-168	Track 24 Cody Lyons	2:46	2007
8.	The Best of Hilltop Meditation	CDSC-122	Track 19 Cody Lyons	3:25	2009

(Hilltop Records,1680 No. Vine St, Suite 321, Hollywood, CA 90028)

CONFLICT IN HISTORY

INTRODUCTION

The hypothesis of 1971 "Conflict in History" by Dennis G. Collins is that history is restricted to certain frequencies (the day, month, year and so forth), which limits what can be calculated (p.107). This idea is similar to the fact that if there are only certain frequencies (colors or spatial frequencies) available for a painting, only certain paintings can result.

Over the past forty years, the author has worked out various cases, some of which are included in this volume. The studies are included in reverse chronological order to when worked on. Attempts by the author to publish over these years have closely paralleled his departure from various universities. R.K. Webb (p.100) said "I do not know any reputable historian who would accept even the possibility of a valid cyclical theory of history." I wanted to asked him if he expected the sun to come tomorrow, and the next day, but I decided to let it go, and in any case he has now passed away I believe.

My attempt to publish again is related to the problem that time is getting short for humanity to learn how to resolve conflicts, such as studied in these pages. The theory permits forecasts of a sort, similar to weather forecasts, which do not seem to bother people, in spite of Biblical warnings against soothsayers (Deut. 18:10). In fact such weather forecasts may be said to save lives on a daily basis.

Certainly the daily cycle, giving rise to day and night, is considered known due to the Earth's rotation on its axis of about 24 hours, although it can effectively be modified somewhat due to artificial light. The monthly (27 day) cycle p.122 is less well understood, although it is related to the moon's cycles and the (nonuniform) rotation of the sun at different latitudes. The yearly cycle is related to the rotation of the Earth about the Sum, giving rise to seasons. The magnetic sunspot cycle of 22.4 years (p.122) is related to the 25-year generational cycle of father to son.

The author called up the Adler Planetarium in Chicago in 1971 to find out if there were a natural 500-year cycle (cf. pp. 120-122) of ages. The personnel said they didn't know of any. However the 2004 *Book of Hiram* (pp. 226-229) by Christopher Knight and Robert Lomas discusses a cycle of planets Venus and Mercury which "appeared to resynchronize every 480 years," related to the so-called "Shekinah."

Organized frequencies can be synthesized by the methods of wavelets, which however compare to the metric system, versus the irregular frequencies of the "English system" of measurements. In the history case, it seems humanity is more-or-less stuck with the irregular frequencies listed above.

Over the forty years the role of the 1.16 expansion factor discussed on p. 193 of the *Time-Dispersed Warp* paper vis-à-vis the Fliess-Swoboda density, has remained somewhat unclear. The ratios 25/22.4= 1.116 and 29.5/27= 1.092 and 500/480 = 1.041 compare to what R.

Buckminster Fuller calls the synergetics conversion factor 3/2.828=1.0606 (p.590 *Synergetics* by Buckminster Fuller 1975).

Limited attempts to work with higher-valued Hermite polynomials Hn (than n=2) in the heterodyned wavelet Exp(-s^2*t^2/2)*H2(s*t)*cos(w*t) with s=3/4 and w =2*Pi and H2(s*t) = 4*(s*t)^2-2 have not proved valuable so far, although if everyone worked with the H2(x) wavelet, it might kick the theory to a higher level, such as H4(x). The H2 may be related to the n=2 derivative in Newton's second law.

Where more than one frequency is involved, as in p. 9, it has to replace w = 2*Pi. For example, bringing in both yearly and monthly frequencies requires taking w = 2*Pi* 13.5 for the monthly wavelet, based on 13.5 cycles of 27 days per year (13.5*27 = 364.5).

There may be a quantum-field-theory version behind the formulas, the closest to which at present seems to be *My Double Unveiled* (2001) by Giuseppe Vitiello. This work has some connection with David Bohm's theories.

The versions of Rupert Sheldrake's *A New Science of Life* (1981) with "morphic resonance" are somewhat consistent with the theory described here.

There is some overlap with H.T. Odum's emergy (= energy memory) theory. The author acknowledges his discussions with H.T. Odum (Sept. 1, 1924-Sept. 11, 2002) on the emergy theory and also the effort spent by his undergraduate history professor Willis Boyd (April 30, 1924-Oct. 17, 1993) to understand the theory.

The source of the "future points" (approximately +-.4, p.92) as well as esp aspects of the theory require more study, although the derivative theory that esp tends to occur when the derivative of the wave function has a maximum has seemed roughly correct.

Mayaguez, PR April 11, 2011

CONFLICT IN HISTORY STUDY OF WORLD WAR II

By Dennis G. Collins, 7108 Grand Blvd., Hobart, IN 46342-6628 or Urb. Mayaguez Terrace, 6009 Calle R. Martinez Torrez, Mayaguez, PR 00682-6630

This analysis of World War II is based on the author's 1971 paper "Conflict in History," which graphs "warp" above the time axis as favorable to one side, and "warp" below the axis as favorable to the other sidei n a conflict, according to the formula: f(x) = exp(-(.75*t)^2/2)*(cos(2*Pi*t)*(4*(.75*t)^2-2)).

At the important +/- two period points, in this case plotted favorable to the Axis powers (mostly Germany and Japan), are the bombing of Pearl Harbor by the Japanese on Dec.7, 1941 and the death of Gen. George S. Patton on Dec. 21, 1945, after an auto accident a few weeks earlier (Please see Figure 1.). Of course the bombing of Pearl Harbor brought the United States into World War II on the side of the British and Russians, mostly forming the Allies (although the Soviet Union had a peace pact with Japan). The Central decision point favorable to the Allies at t=0 around Dec.1943 saw the Russians break the German siege of Leningrad (now St. Petersburg again) on Jan.27, 1944 and the U.S. start its "island hopping" campaign toward Japan with attacks in Tarawa (Nov.20, 1943) and Kwajalein (Jan.31, 1944).

The supposed one-year advance of the Allies June 1944 to June 1945 is easy to spot as dating from the Eisenhower-directed D-Day invasion of Normandy (France) June 6, 1944 to include the invasion and surrender of Germany May 7, 1945, and the development of the atomic bomb for use against Japan. The actual dropping of the bombs in August 1945 and victory in the war against Japan was still in the period favorable to the Allies, after a year of advances in the Pacific under Douglas MacArthur. Curiously the Battle of the Bulge, Dec.1944 appears as a bulge on the graph, favorable to the Axis, about halfway through the one-year advance of the Allies.

The supposed one-year advance of the Axis powers from June 1942 to June 1943, is more difficult to justify, due to blunders and/or lack of intelligence by the Axis. However there were several mostly ill-fated offensives by the Axis powers during this period. Rommel's African campaign, with the capture of Tobruk June 21, 1942 resulted in failure to secure oil supplies and the expulsion of Axis forces from Africa by May 12, 1943, although it did lead to further German occupation in southern France. A renewed German offensive in Russia resulted in German troops capturing Sevastopol on the Black Sea in May 1942 and entering Stalingrad (now Volgograd) on Sept. 16, 1942; however a Russian counterattack Nov.19, 1942 resulted in complete surrender of German forces under von Paulus, Jan. 1943 around Stalingrad. Nov.1942 saw the maximum Allied shipping loss due to German submarine "wolf packs," with as many as 235 U-boats in action in spring 1943. Japanese attempts to maintain their Pacific empire fell apart with the shooting down (again due apparently to intelligence failure) on April 18, 1943 of Admiral Yamamoto, who planned the Pearl Harbor attack (For example organized Japanese resistance ended on Attu in the Aleutians on May 30, 1943.), and the Japanese retreated across the Yangtze in March 1943, although Japanese continued to penetrate the south Asian continent and even entered India March 22, 1944 at a period favorable to the Axis. The ending of the "one-year advance" saw the Allies invade Sicily July 10, 1943 and land in Italy Sept. 3, 1943 moving toward the decision point Dec.1943. Jan 1944 saw the battle at Anzio with the fall of Rome June 4, 1944 inaugurating the one-year Allied advance.

Supposed Allied high point May 1940 saw the Germans fail to destroy the British/free French army at Dunkirk and a supposed turning point in the air Battle of Britain through summer 1940, which give Britain air superiority at least during the day; high point for the Allies around June 1941 saw Lend-Lease take effect March 11, 1941 and the embargo against Japan. High point around June 1942 saw the Battle of Midway June 4-6, 1942 destroy most of the Japanese carrier fleet as a turning point similar to the Battle of Britain, and the death of the so-called Nazi "Hangman" Heydrich June 4, 1942.

Other points favorable to the Axis saw Germany free from attack during the winter of 1939-40 after conquering Poland, starting Sept. 1, 1939 at the beginning of World War II, and launching the "Blitz" night air attacks against Britain Sept. 7, 1940 to May 6, 1941. Also the period favorable to the Axis saw the first V1 guided missile attack against London June 13, 1944.

A couple events contradictory to the theory include the sweeping German occupation of Europe summer 1940, including the fall of France June 22, 1940, and the death of U.S. Pres. Franklin D. Roosevelt on April 12, 1945 at a supposed high point of the Allied cause. Again these events may become clearer at a one-month level of analysis; for example both Mussolini (April 28, 1945) and Hitler (April 30, 1945) died approximately 2 weeks (or ½ period) after Roosevelt at the monthly level.

Also a preliminary study of the "future" points at July-Aug. 1942 and April-May 1943 turns up only two items: the Allied invasion of Sicily in July 1942 and the Allied invasion of Hollandia (now Jayapura) on the island of New Guinea, so that more study is needed. Hopefully it doesn't indicate more encounters with mafia and cannibal situations.

REFERENCES

World War II, *The World Book Encyclopedia*, Chicago, IL 1974.

Individual Micropedia articles, *The NewEncyclopedia Britannica*, Chicago, IL 1990.

List of Thermodynamic Modeling papers by Dennis G. Collins (for INFORMS Conference July 11, 2007, Wyndham Rio Mar, Puerto Rico)

Dept. of Math Sciences, Univ. of Puerto Rico,
Mayaguez, Box 9018, Mayaguez, PR 00681-9018

A Thermodynamic Goal Model (1978)

A Version of Thermodynamics Applied to a Social Science Problem: The American Motors Merger (1980)

On Some Analogies of Thermodynamics of Thermodynamics with Problems in Social Science (1981)

Liquid-Vapor Models (American-Braniff Airlines) (1983)

Oil—Merger Vs. War in Iran and Iraq (AMS-Jan 9, 1985, Anaheim, CA)

Liquid-Vapor Models for Social Science Problems (AMS-Jan 9, 1986, New Orleans, LA)

Thermodynamic Modeling of US/USSR Relations (April 12, 1985, Eastern Michigan Univ., Ypsilanti)

Liquid-Vapor Modeling of Merged Components for Social Science Problems (US/Japan WWII) (AMS, Jan 21, 1987, San Antonio, TX)

A Thermodynamic Setting for the Phillips Curve (*Mathematical and Computer Modeling, Vol.14, pp. 1183-1188*) (19990)

A Case Study of Puerto Rican Wage Level (P) Versus Unemployment (U) for 1980-1988 by Thermodynamic Modeling (Feb 24, 1990, Arecibo, PR)

A Thermodynamic Version of the Phillips Curve (August 9, 1990, Baden-Baden, (West) Germany)

Remarks on the Gulf Crisis (Jan 17, 1991)

Measurement of Temperature in Economic Systems Considered as Thermodynamic Models (July 1991)

On the Dangers of Downsizing from a Thermodynamic Modeling Viewpoint (June 23, 1994, World Bank, Washington, DC)

On the Thermodynamics of Mime-Matter Interactions (April 20, 1996, Arecibo, PR)(June 7, 1996, Kalamazoo, MI)(July 10, 1996, Gainesville, FL)(May 3, 1997, Deroit, MI)

Business-Generalized Thermodynamics (April 12, 1997)

On the Generalized Thermodynamics of Mime-Matter Interactions (1997)

Temperature-Based Ascendancy Derived from a Cost or Reward Function (Aug 4, 1999, Boise, ID MAXENT)

IPO's and the Vapor-Pressure Curve (July 21, 2000, Toronto, Canada ISSS)

Thermodynamic Modeling of Moral Codes (=Moral Codes I) (Nov 16, 2000, Mayaguez, PR) (Feb 24, 2001 Humacao, PR)

SigmaXi poster (Nov 10, 2001, Raleigh, NC, SigmaXi Annual Meeting, discussion one year later 2002)

Moral Codes II (Nov 2003)

Divvy Economies Based on (an Abstract) Temperature (Aug 2003, Jackson Hole, Wyoming MAXENT)

"Tropical" Emergy and (Dis-)Order

Dennis G. Collins, Dept. of Mathematics, U. of Puerto Rico, Mayaguez

ABSTRACT

The title does not refer to emergy calculated for the tropics, but to simplified math systems that still retain some of the features of the full system. "Tropical" Emergy refers to a version of emergy that may be calculated for cubes located on, say, a two-dimensional or three-dimensional grid. The cubes can be considered as chairs placed in a room. The number of exposed lateral or side surfaces of the cubes or chairs can be considered as a simplified measure of gradients or Fisher information, which are, in turn, a substitute for energy. For example the thirty-foot waves of a hurricane with large near-vertical gradients have more energy than a near-calm sea. If two cubes are placed together, two side surfaces disappear, reducing "tropical" Fisher information of the two cubes. Emergy can be defined as related to the amount of tropical Fisher information used up in creating a structure. For example, if sixteen random cubes are combined into four groups of four two-by-two squares, the number of exposed surfaces decreases from 16x4 = 64 to 4x8= 32, leading to a transformity of 64/32=2. Examples are worked out and other types of order discussed.

INTRODUCTION

Since H.T.Odum was always interested in explaining his work to young audiences, for example by the use of icons that did not require reading ability, to develop network theory, he would have embraced the idea of studying simpler systems that convey some of the properties of emergy algebra, with only limited technical details. This procedure is similar to studying algebraic systems that have only some properties of the usual structure. For example matrix multiplication is not commutative (ab is not necessarily ba), whereas typical multiplication of real numbers is.

Such so-called "tropical" information measures were outlined in the author's paper [Collins, 2005]. Readers interested in "tropical" algebra as a math subject can consult papers by Berndt Sturmfels and others. For these simple models, quantities such as emergy and transformity can often be calculated exactly, and various choices studied more easily as to their effects. In the previous paper [Collins, 2005] three types of order were discussed: 1) clumping order (measured by adjacent surfaces used up or decrease of Fisher information), 2) stacking order (measured by decrease of entropy), and symmetric order. A questionnaire found no agreement as to which configuration had the most order: A) stacking five chairs or cubes on top of each other, B) placing five chairs adjacent in a row, C) placing three chairs adjacent in a row and two chairs symmetrically on top of the three chairs, D) placing the chairs around a room like the five-pattern on a set of dice (somewhat like chairs in a classroom) and E) placing the chairs "randomly." As mentioned in the ABSTRACT, emergy seems most related to clumping order.

However in a weightless environment, there would be no distinction between clumping and stacking order, as adjacent cubes would become stacked if rotated 90 degrees from horizontal to vertical. Thus the top and bottom surfaces would be considered equivalent to lateral surfaces, which effectively combines energy and entropy calculations together. However it must be observed that this procedure "goes the same way" as official entropy calculation, i.e. stacking blocks reduces the number of exposed top and bottom surfaces, thus decreasing our result, just as it decreases official entropy.

In the following sections four levels of C (or cell) transformity are considered, applying direct calculation and the program presented in the 1st emergy conference [Collins and Odum, 2000]: a) block, b) wall, c) tower, and d) castle. The goal is to explain how transformity works, as well as the program that calculates it. It is found that the procedure covers only reversible energy, so that the C-transformities come out rather small, but by including feedback loops, irreversible energy may be included to yield a more appropriate transformity measures. Typically symmetric order is obtained by irreversible energy, as the energy spent straightening a picture frame cannot be recovered, but the energy spent stacking a chair can be recovered to some extent. It is not entirely clear if there is a term within thermodynamics to refer to the type of energy considered in this paper; however it may reduce to a type of internal energy, or embodied energy if considered as sub-type of energy.

BLOCK TRANSFORMITY

Blocks can be created as combinations of cubical cells, or the "chairs" discussed in the previous section. This process is completed by companies such as the LEGO company. The creation of blocks from one-by-one cells is similar to the embodiment of solar energy by photosynthesis. Thus the one-by-one cell plays the role of the solar emjoule. Basically nothing can be built from one-by-one cells since there is no way to stick them together laterally (although they can be stacked on top of each other, which is not too interesting).

Although actually LEGO blocks are constructed at one-third and other heights, here they are all considered to be the same height. With this restriction, no top/bottom surfaces are used up in the block construction process. Also only double-wide blocks are considered from now on.

C-transformity is calculated as IN/OUT or total original exposed lateral plus top/bottom surfaces divided by total final exposed lateral plus top/bottom surfaces. Please see Table 1 for the results. This type of transformity would be created by the brick factory.

PROTO-WALL TRANSFORMITY

Building things with the blocks is made possible by putting prongs on the top surfaces and receiving grid holes on the bottom surfaces. Building something with the blocks requires at least two levels, i.e. stacking or entropy decrease, if it's kept track of. A perhaps overstrict stability criterion is that there is no vertical line of height two between blocks. With this restriction, various structures of height two, here called proto-walls, can be built and their C-transformity calculated. Three proto-wall structures are presented in TABLE 2.: 1) a 6x2x2 complete proto-wall made out of two 4x2 blocks and two 2x2 blocks, 2) a 10x2x2 incomplete proto-wall made out of four 4x2 blocks in a kind of "Z" pattern, and 3) a 12x2x2 complete proto-wall made out of five 4x2 blocks and two 2x2 blocks. As mentioned above only double wide walls can be constructed wth double-wide blocks.

Of course, many more proto-wall types can be built. However, their basic C-transformity seems to be in the range of 2 to 3. Any combination of blocks that gives the same result will calculate to the same C-transformity, although not all would be stable. Later a way to add to C-transformity through feedback to take stability into effect will be presented. Proto-walls can be stacked on top of each other and their C-transformity calculated; this case is left as an exercise.

TOWER TRANSFORMITY

Four proto-walls can be connected to form a proto-tower, i.e. a tower of height two. A rather striking result is that any rectangular proto-tower (i.e. with height 2) has C-transformity exactly 3. The general result for the C-transformity of a tower of height m is TR= 6m/(m+2), which reduces to 3 in case m = 2. This value 3 is also the limiting C-ransformity of an infinitely long proto-wall. The limiting C—transformity for an infinitely high tower is 6 as m goes to infinity.

CASTLE TRANSFORMITY

By combining 3x2 and 4x2 blocks, it is possible to create a stable proto- "castle" with 4 towers and 4 connecting walls (all at height 2). In the case of 12x12x2 proto-towers and 6x2x2 proto-walls, the result has C-transformity 3.067 due to the surfaces used up connecting walls to the towers.

All of the above structures can increase C-transformity by going from a "proto"-structure with height 2 to a structure of height m. For example, a "castle" of height m=6, made by placing 3 proto-castles on top of each other, increases C-transformity to 4.651, with believed limit 6 as m goes to infinity.

TRANSFORMITY PROGRAM

The program of the 1st emergy conference [Collins and Odum, 2000] seems to work well to calculate C-transformity, as Figure 1 illustrates, although it cannot distinguish stability. For example as well as combining 3x2 and 4x2 blocks, it is possible to make a proto-castle as described above by combining 4 wall of type 6x2x2 with 4 proto-towers of size 12x12x2, although the result would not be stable because there is nothing to stick the walls to the towers. However the C-transformity comes out the same. Please see Figure 1. It may be mentioned that the rows of a consistent program matrix can be multiplied by any factor without changing transformity results. However if everything is done on the same scale (for example in terms of exposed cell surfaces), it is easy to build up new structures from previous rows and calculate their transformity. Thus it is possible to construct general schema in terms of number of structures combined and their result, which can be put in as rows of the program matrix. It is only necessary to include rows whose output is needed later.

EMERGY VERSUS "INTERNAL" ENERGY

So far, increasing C-transformity of the block calculations seems to go with increasing complexity. However, as mentioned above, the technique employed here seems to measure a type of internal energy versus all the energy used up to make something, i.e. it is possible to distinguish between reversible energy used up to make something, which can be recovered; and irreversible energy (which cannot be recovered). H.T. Odum stressed the irreversible energy losses at each stage as contributing to emergy. Here for example, no attempt is made to calculate the energy required to move the blocks around, which, if included, would greatly increase transformity.

Further, for example, including windows and doorways in the wall and castle structure would expose more surfaces and thus decrease C-transformity due to embodied energy, while apparently increasing complexity. Typically such structures as doorways require much more effort and thus lost energy and would overall increase emergy. Of course towers and castles with no doors are effectively useless.

As another example, constructing a solid cube with the blocks, disregarding stability considerations, can increase C-transformity to infinity, although it does not appear to increase complexity. The C-transformity of a solid cube goes up because all the interior surfaces are used up.

FEEDBACK

Here is where feedback comes to the rescue. The program so far seems to take into account bottom up or feed forward considerations rather well. However one of the criticisms of Odum's network theory was lack of ability to take into account feedback structure. That this criticism is unwarranted can be seen through the ability of feedback to take into account irreversible energy losses. These losses appear in the upper right triangle of the program matrix. For example the energy required to put the blocks together laterally to form a wall can be considered as irreversible energy due to a high-transformity mason. And the energy required to tell the mason where to put the wall (i.e. the sound energy of the words "Put the wall here.") can be considered the even lesser energy input (called "control information" in [Corning and Kline, 1998]) due to an even higher-transformity architect. Please see Figure 2. A way to distinguish between stable and unstable wall structures is through the top-down feedback energy of an expert who supervises the stable construction, thereby increasing its transformity or energy quality. Of course the expert would be responsible for including the proper ingredients. As expressed in [Collins, 2005], order appears to be substantially in the eyes of the beholder, so that various inputs of energy, even advertisement money, can be considered to increase transformity or emergy. This view is opposed to Marx's theory of value, which only considers work of production as contributing value.

CONCLUSION

This paper attempts to explain networks and transformity by a simpler calculation of cell or "C-transformity" that can be done exactly: how many exposed surfaces to start out with, supposing cells scattered randomly and how many exposed at the end. This reversible or embodied (internal) energy can be recovered if the cells are separated, exposing their surfaces again. The transformity program shows how this bottom-up calculation can be carried out step-by-step without going back to the original cell-level. Feedback allows things to obtain higher transformity through substantially irreversible top-down energy inputs. It is hoped these ideas can apply to evolutionary development as outlined in [Carroll,2005], as well as advance H.T. Odum's ideas. As a reviewer pointed out this paper is limited in that it does not cover deeper topics of emergy algebra, such as splits and co-products. Perhaps further study will reveal analogs of these properties in the tropical emergy.

ACKNOWLEDGMENTS

The author thanks high school senior Mr. Humberto Diaz for helping with some early calculations, including a more-complicated tower-type structure which also had a doorway and roof. As explained above, even though the doorway decreases C-transformity, it may increase overall transformity through top-down energy inputs by experts on where to put the door and how to build it, and so on. Also the author thanks high school senior Mr. Glenn Collins for help on computer problems. The author also thanks the College of Arts and Sciences at the University of Puerto Rico, Mayaguez for a four-hundred-dollar travel-money grant to attend the 4th Emergy Research Conference.

Remark Added. A topic of concern at the 4th Emergy Research Conference was the interpretation of emergy maximization. Perhaps Prof. Corrado's writing emergy increase as a product of dissipative and generative emergy can throw light on this problem, since the problem becomes similar to the standard calculus problem of optimizing the area (=length*width) of a rectangle with given perimeter, e.g. if dissipative emergy (=length) is large as in a wasteful fuel process but generative emergy (=width) is small due to little innovation, the overall product will remain small, or if a process is highly innovative (large width) but has very limited scope (say only milligrams can be produced) the overall product will also remain small, but maximum emergy is found in an in-between method, and so on.

REFERENCES

Carroll, Sean B., 2005. *Endless Forms Most Beautiful*, W.W.Norton, New York.

Collins, Dennis, 2005. Algorithm for Minimum Laterally-Adiabatically Reduced Fisher Information, *Bayesian Inference and Maximum Entropy Methods in Science and Engineering*, 25th *Int. Workshop, San Jose, CA, AIP Conf. Proceedings 805*, ed. Kevin Knuth et alia, American Institute of Physics, Melville, NY, p. 345-352.

Collins, Dennis and H.T. Odum, 2000, Calculating Transformities with an Eigenvector Method, *Emergy Synthesis*, ed. Mark Brown et alia, Center for Environmental Policy, Gainesville, FL, pp.265-280.

Corning, Peter and S.J. Kline, 1998. Thermodynamics, Information and Life Revisited, Part II: Thermoeconomics and Control Information, *Systems Research and Behavioral Science*, 15: 453-482.

TABLE 1. Block Transformity (TR) or C-Transformity

Dimension	Lateral	Top/B	Total IN	Lateral	Top/B	Total OUT	TR=IN/OUT
1x1 block	4	2	6	4	2	6	1
2x1=2 cells	8	4	12	6	4	10	1.2
3x1=3 cells	12	6	18	8	6	14	1.285
4x1=4 cells	16	8	24	10	8	18	1.333
2x2=4 cells	16	8	24	8	8	16	1.5
3x2=6 cells	24	12	36	10	12	22	1.636
4x2=8 cells	32	16	48	12	16	28	1.714

TABLE 2. Proto-Wall C-Transformity

Observe in Figure 1: The program can calculate proto-wall transformities in terms of block results instead of cell results, and so on.

Dimension	Lateral	Top/B	Total IN	Lateral	Top/B	Total OUT	TR=IN/OUT
6x2x2 complete proto-wall							
2 2x2's =8 cells	32	16	48				
2 4x2's=16 cells	64	32	96				
Total 24	96	48	144	32	24	56	2.571
10x2x2 incomplete proto wall							
4 4x2's=32 cells	128	64	192	40	40	80	2.4
12x2x2 complete proto wall							
2 2x2's=8 cells	32	16	48				
5 4x2's=40 cells	160	80	240				
Total 48	192	96	288	56	48	104	2.769

Program Matrix

Cell	2x2	3x2	4x2	6-wall	10-wall	12-wall	12tower	pr-castle	6-high	castle
24	-16	0	0	0	0	0	0	0	0	2x2 from cells
36	0	-22	0	0	0	0	0	0	0	3x2 from cellls
48	0	0	-28	0	0	0	0	0	0	4x2 from cells
0	32	0	56	-56	0	0	0	0	0	fr 2 2x2 & 2 4x2
0	0	0	112	0	-80	0	0	0	0	from 4 4x2 block
0	32	0	140	0	0	-104	0	0	0	fr 2 2x2 & 5 4x2
0	0	0	560	0	0	0	-320	0	0	tower fr 4x2
0	0	352	2240	0	0	0	0	-1440	0	from 3x2 & 4x2
0	0	0	0	224	0	0	1280	-1440	0	fr 4 wall & tower
0	0	0	0	0	0	0	0	4320	-2848	from 3 pr castle

TR C-transformity output

Cell	1
2x2 block	1.5
3x2 block	1.636
4x2 block	1.714
6x2x2 cl pr-wall	2.571
10x2x2 incl pr-wall	2.4
12x2x2 cl pr-wall	2.769
12x12 pr tower	3
pr castle	3.067
6-high castle	4.651

FIGURE 1. The point of the program is that (C-) transformity can be calculated from numbers of later ingredients instead of directly from cells. Both stable and unstable ways of creating proto-castles have the same C-transformity 3.067 (2nd and 3rd rows from bottom).

Program Matrix

Cell	2x2	mason	architect
24	-16	0.002	0.0002
10000	0	-1	0
100000	0	0	-1

TR Transformity output

Cell	1
2x2 block	4
mason	10000
architect	100000

FIGURE 2. Smaller and smaller feedback input energy from higher and higher transformity agents contributes to the overall transformity. The basic C-transformity 1.5 of the 2x2 block is increased to 2.75 by the mason, say putting the block in an actual straight line (2.75 found by putting architect feedback .0002 to 0), and then to 4 by the architect marking out the straight-line position for the block. The last two types of energy input may be considered as increasing symmetric order. Another type of order may be taken as functional order, which has to do with getting things to work, or arranging them in time sequence, in a fourth dimension.

CONFLICT IN HISTORY STUDY OF THE AMERICAN REVOLUTION

By Dennis Collins, 7108 Grand Blvd., Hobart, IN 46342-6628 or Urb. Mayaguez Terrace, 6009 Calle R. Martinez Torres, Mayaguez, PR 00682-6630

This analysis of the American Revolution or War for Independence is based on the author's 1971 paper "Conflict in History." At the important +/- two period points, two dates stand out: British defeats of Gen. Burgoyne at the battle of Saratoga, Sept 17-Oct. 17, 1777 and of Gen. Cornwallis at the battle of Yorktown, Oct. 6-19,1781. Please see Figure 1. The first victory brought France into the War on the American side and the second led to American independence, confirmed by the Treaty of Paris, ratified April 19, 1783. The central decision point Oct. 9, 1779 saw Count Pulaski wounded in a siege of Savannah, Georgia, leading to his death two days later, which reflected the inability of the Colonials to gain control of all British settlements in the North American continent, especially Canada. In general, the "decision point" around Oct. 1779 seems to have been a disaster for the Indians, with the "Six Nations" or Iroquois Confederation dating from 1100 to 1500 A.D. being broken and scattered. For example the back-and-forth raiding and retaliation seems to have reduced the population of the Mohawk Valley in New York from 10,000 down to 3,000 from 1777 to 1781, technically a victory for a "scorched earth" British policy. *Gerald Horton's Historical Articles* on "What happened to 7,000 People?" says, "The winter of 1779/1780 was the worst possible time for the Indians to have lost their homes and crops," it being the "Hard Winter" of the "Little Ice Age," during which winter, "hundreds of Indians died of disease, exposure, or malnutrition."

The supposed one-year period of steady advances of the Colonial side ran from April 1778 to April 1779, with Congress ratifying the treaty with France May 4, 1778 (signed by King Louis XVI on Feb. 6, 1778), forcing the British to a standstill in the North, and a British surrender at Vincennes, IN on Feb. 25, 1779 to the forces of Lt. Col. George Rogers Clark in the West. However Clark did not have sufficient men to reach Detroit, leaving Canada as a safe haven for the British at the supposed decision point Oct. 1779.

The British turned their efforts to the South and had a supposed one-year period of advances from April 1780 to April 1781. On May 12, 1780 the Americans suffered the worst defeat of the war when the British captured Charleston, South Carolina and the entire Southern American Army. The battle of Camden, South Carolina, Aug. 16, 1780 saw most of the army of the Colonial Gen. Gates destroyed. As late as June 3, 1781, Cornwallis' men under Tarleton raided Jefferson's estate at Monticello, Virginia, although Jefferson escaped and had moved his family earlier. Stalemate mostly persisted in the North, in spite of problems for the Colonials. On May 25, 1780, after enduring cold and depravation during their winter at Morristown, New Jersey, two regiments of Washington's army threatened mutiny. Troops were dispatched from Philadelphia to put down the rebellion and the two leaders of the mutiny were hanged. Benedict Arnold's treachery was discovered Sept. 23, 1780, although he escaped to join the British. Future President Andrew Jackson was captured by the British at age 13 on Aug.6, 1780, together with his brother Robert, who died in captivity. Andrew was released through the efforts of his mother, who died March 15, 1781 after nursing Andrew back to health and attempting to treat other Colonial casualties.

Supposed "future" points occurred around May-June 1779 and Feb.-March 1780. Somewhat limited research indicated that in the North there was a shortage of goods and inflation around June 1779 with a letter of alarm from the Boston Committee of

Correspondence to the Continental Congress in Philadelphia on June 21, 1779. Also Spain declared war on Britain June 21, 1779. Tina Nelson of the Baltimore County Public Schools writes: "British commanders formulated an official policy toward the use of slaves in June, 1779," which allowed masters to hire out slaves to the British army. Apparently Colonials "ran their slaves" away from battle areas to prevent any from fighting for the British, and many slaves that did try to join the British had a fate similar to the Indians. May 1779 saw Benedict Arnold beginning secret negotiations with the British around New York. A curiosity [Wead, p.431] is that future Pres. Andrew Jackson's older brother Hugh Jackson was killed May 29, 1779 fighting for the Colonials. At the second "future" point, on March 22, 1780 there was a British raid of Col. Duncan MacPherson with 300 of the Black Watch outside their enclave in Manhattan, N.Y. across the river into the "Neutral Ground" of New Jersey around Hackensack, apparently with little consequence except pillaging, burning and looting before being forced back. Perhaps these mostly lackluster events at the so-called "future points," indicate a mostly peaceful future, or at least a standoff coming out of the Revolutionary War. Of course, there is the question of how far in the future the "future points" are. One answer is that the "future points" are like radiation leaking out of a potential well, which repeats periodically at a given frequency. Another view is that "future point" events are sort of cancelled by specific events in the future. Something like the pattern of the March 22, 1780 raid is repeated in the British raid on Washington, D.C. in the War of 1812. A combined theory might give repetitions with differing amplitudes or outcomes. Perhaps further research, such as studying the papers of Washington and Jefferson, will turn up other events or theories.

Besides above-mentioned high and low points, the theory predicts events favorable to the American cause around Aug. 1775 and Sept. 1776, as well as Nov. 1782 and Dec. 1783, according to the periods being somewhat more than one year. July 3, 1775 saw George Washington take command of the army, and of course July 4, 1776 saw the signing of the Declaration of Independence. The early high points for the Americans may not represent American victories, so much as the failure of the British to finish off the Colonial cause. The occupation of New York City by the British on Sept. 15, 1776 may be seen as contradictory to the theory; thus these events may perhaps be studied in more detailed monthly cycles, as Washington's victory at Trenton, Dec. 26, 1776. November 30, 1782 saw the American and British sign a preliminary peace treaty in Paris granting independence, with the final peace treaty signed Sept. 3, 1783 in Paris, France. Dec. 14, 1782 the British left Charleston and Nov. 25, 1783 the British left New York City.

Supposed low points for the Colonial cause or points favorable to the British according to the theory come about Feb. 1775, March 1776, as well as May 1782 and June 1783. The death of Jefferson's Mother March 31, 1776 and his wife Martha due to a difficult childbirth May 20, 1782 (although lingering on to Sept. 6, 1782) sort of neatly bracket the war years as seen by the theory. Again the events may have to be studied in more (monthly) detail, as for example the evacuation of Boston March 17, 1776 by the British may be seen as contradictory to it being a high point for the British.

Thus further topics for study remain. 1-15-2008

REFERENCES

The Revolutionary War in America, *The World Book Encyclopedia* (main reference).

Thomas Jefferson and His World, by Editors of *American Heritage* (*Junior Library*, narrative Henry Moscow), Troll Associates, Mahwah, NJ, 1960.

Wead, Douglas, *The Raising of a President*, Atria Books, New York, 2005.

The Neutral Ground of New Jersey during the Revolution, http://www.doublegv.com/ggv/battles/Neutral.html.

The Revolutionary War, http://home.hiwaay.net/~jpkilpat/amerrev.htm.

Horton, Gerald, What Happened to 7,000 People?, http://hortonsarticles.org/hh7thousand.htm.

Ottman, Rachel M. (5th grade teacher, McCarthy Middle School, Chelmsford, MA), Abigail's War, Massachusetts Historical Society 2004, http://www.masshist.org/digitaladams/aea/index.html.

The American Revolution, http://www.bright.net/~double/rev.htm.

Nelson, Tina, Fighting for Whose Freedom? Black Soldiers in the Revolution, Baltimore County Public Schools, internet article.

SOME STUDY PROBLEMS OF THE QUANTUM HISTORY INSTITUTE

By Dennis Glenn Collins 9-20-2011

1) Nuclear War—Edge of the Cliff. Gradually the people involved with and who remembered the U.S. Compromise of 1820 to preserve the Union died off, and in fact the Compromise was effectively repealed in the 1850's, with the result that the United States slid over the cliff into the Civil War. Currently the people who remember the Cuban Missile Crisis of 1962 are dying off, with the danger of the world sliding over the cliff into nuclear war. In fact many of the Cold War and Israeli weapons may be becoming unstable due to rust and deterioration of wires gauges and so on.

2) Conflict in History—Morphic Fields. The problem is to explain how the conflict in history theory arises from frequency restrictions, by quantum field theory or other means. The background is given in *Some Conflict in History Documents* by Dennis Collins (2011). Relevant book is *My Double Unveiled* by G. Vitiello and work by Rupert Sheldrake.

3) Economic Crises of 2008 to present—Thermodynamic Modeling Update. The problem is to extend the Gibbs thermodynamic models to explain the current economic crises and tell how to solve them. The background is given in "Merger Vs. War in Iran and Iraq" and other papers by Dennis Collins. Relevant book is *On the Brink* by former U.S. Treasury Sec. Henry Paulson.

4) Group theory symmetry—Symmetry patent view. The problem is to extend the Collins symmetry patent U.S. #7,873,220 B1 (Jan 18, 2011) to discrete and continuous group theory and semi-groups. A relevant beginning paper is "Measuring the Symmetry of a Finite group" by Dennis Collins (2011). Krohn-Rhodes Theorem may be relevant to semi-groups.

5) Entropic Gravity—Symmetry view and practice. The problem is to explain gravity in terms of entropy and symmetry, and apply results to practical issues such as obtaining weightlessness by coordinating or disorienting electron or atomic configurations. A relevant paper is "On the Origin of Gravity and the Laws of Newton" by Erik Verlinde and "Symmetry of the Wedge and other shapes" by Dennis Collins. There is possible relevant work by Ariel Caticha.

6) DNA biology—Genetic Firemen. The problem is to figure out what to do if DNA modification gets out of control and can't be contained. Some relevant work is creation of new organism by CraigVenter.

7) Cancer Research—symmetry viewpoint and/or political spectrum model. The problem is to model and detect cancer by symmetry method. A relevant paper is

"Detecting some 2-D malignant tumors by 1-D continuous symmetry" by Dennis Collins. The political spectrum model is based on cancer as a tyranny model intermediate between bone and soft tissue, creating a "mini-country." A relevant paper is "Political Spectrum Models" by Dennis Collins and David Scienceman.

8) Bottle timer—Planck's constant calculation. The problem is to calculate Planck's constant from the properties of space. The bottle timer, a quantum version of the hourglass, with patent U.S. # 6,533,450 B1 of Glenn and Dennis Collins (2003) may provide a key to this project. Relevant work is by Dewey Larson and later followup, and models by James Clerk Maxwell, as in *On Physical Lines of Force*, 1861 (end of Part II, after p.348, p.38).

9) Hurricane, tornado earthquake models—political weather modification. The problem is to develop tractable models of hurricanes and tornados, as well as earthquakes. Of course many models already exist. Another problem is to study modification of weather by human activity such as global warming or wind turbine friction and/or political manipulation of weather by satellites or electrical pumping a la Tesla.

10) Eigenvalue study—save the Union. The problem is to follow up on papers by Dennis Collins, such as SIAM 2002, "M-Matrices and Emergy" and "Eigenvalue Ranking Methods" to help preserve the United States from economic attack. Work by Murray Patterson may be relevant.

11) Solar energy. This problem is a standard one of finding more efficient systems of capturing solar energy that do not set up other instabilities, as of the alleged Atlantis problem of "tuning the crystals too high."

12) Consciousness based on fractal models. In New Orleans, Dennis Collins asked P.A.M. Dirac if he thought consciousness was a real, physical property. Dirac answered that you would have to show where it entered the equations of physics. Possible relevant work is *On the Origin of Consciousness in the Breakdown of the Bicameral Mind* by Julian Jaynes.

13) Political spectrum models. The problem is to study in more detail and expand the paper "Political Spectrum Models" by Dennis Collins and David Scienceman (vol. 5 *Emergy Synthesis*, Gainesville, FL).

14) Symmetry patent—Protein structure. The problem is to enlarge the Dennis Collins algorithm of the patent U.S. #7.873, 220 B1 to cover problems of protein structure, as well as more complicated pattern identification.

15) Inflation model—boundary value problem. The objective is to extend the 20-degree model of Dennis Collins to higher degree or 3 dimensions or orthogonal functions to see if the continuity requirement automatically introduces wave structures.

16) Reflexive structures—2nd order cybernetics. The problem is to model reflexive logical or other structures. Book by Douglas Hofstadler may be relevant, as well as cybernetic journals.

17) Geometric algebra. The problem is to find valid applications or further properties of geometric algebra of David Hestenes.

18) Riemann hypothesis. The problem is to prove or disprove the Riemann hypothesis. Don't expect much progress on this study. Another standard problem is to prove or disprove P = NP complete (and other Clay problems).

19) Language study. The problem is to explain the neurophysiology of language development.

20) World cartoon news. The problem is to develop a cartoon news service that will be the same explaining world events over the whole world.

Besides above is the problem of cataloging books, art, and music of Dennis Collins and his collections.

Toward a Mathematical Origin of Species

By Dennis G. Collins

UPR-Mayaguez (retired)
Urb. Mayaguez Terrace
6009 Calle R. Martinez Torres
Mayaguez, PR 00682-6630
Feb. 26, 2010 for SIDIM-25

ABSTRACT: Adding a new species mathematically involves increasing the dimension of the system of differential equations that describe the interactions of the existing species. Corrado Giannantoni indirectly gave one procedure to do this increase in Emergy Synthesis 5 (2009), pp. 581-598 "From Transformity to Ordinality," based on the emergy algrbra of H.T. Odum. This paper discusses Giannantoni's model and alternatives, although not going into DNA and genome changes.

Introduction

Adding a new species to an ecological web can be done formally by a "BUILDER" pattern, as in computer science (Gamma et al., 1995, p.97). This pattern can open a dialog box in which the properties of the new species are spelled out, as is already done in some simulation models, as for different trees in a forest.

Species can also refer to chemical species and again a procedure can be set up for adding chemical species to an organic or inorganic reaction, which basically increases the dimension of the system of differential equations describing the system.

A more formal method can be devised for quantum theory ("annihilation" and "creation" operators for the harmonic oscillator), and quantum field theory, which involves adding nodes and edges to a graph (Braunstein et al, 2008, p.15) with TPCP's (trace-preserving completely positive maps).

Thus a deeper study of the conditions for adding a species in biology is desired. This process can be carried out at the level of the environment (selection pressures), the phenotype (organism) or genotype (DNA) and leads to topics such as sympatric speciation and allopatric speciation versus concepts of species such as the biological, morphological, mating recognition, cohesion and ecological species classifications. (Wolfe, 2010). This paper concentrates on the ecological level.

114

GIANNANTONI'S APPROACH VIA H.T. ODUM'S EMERGY ALGEBRA

Corrado Giannantoni has attempted indirectly to describe the increase of quality in adding species by H.T. Odum's Emergy Algebra (Giannantoni, 2009). *Emergy* is roughly "energy memory" or energy concentrated into possibly higher forms. The Emergy Algebra covers how emergy changes over ecological pathways. Based on Odum's work (Odum, 1996), Giannantoni recognizes three processes of Creation of Ordinality (p.582):

1) Co-production
2) Interactions
3) Feedback.

Based on some novel math procedures, Giannantoni links the three types of emergy flow to dynamical processes (p.587):

1) Co-production yields "binary" functions, based on fractional derivatives:
   Incipient half-derivative of Exp(a(t)) =+$\sqrt{a(t)}$ Exp(a(t))
   $\qquad\qquad\qquad$ -$\sqrt{a(t)}$ Exp(a(t))

2) Interaction yields "duet" functions :
   Incipient second-derivative of Exp(a(t)) = [a(t)^2, a(t)^2] Exp(a(t))

3) Feedback gives "duet-binary" function:
   Incipient 2/2 derivative of Exp(a(t)) = +a(t), +a(t) $\quad$ Exp(a(t)).
   $\qquad\qquad\qquad$ -a(t), -a(t)

Remark: The italic "*a*" involves "incipient" derivative of a(t). The incipient derivative seems to involve a failed product law of differentiation, whereby (uv)' = u'v (only one term).

Here observe that roughly 1) increases the dimension of the system, i.e. adds a species.
Also 2) creates interaction power terms of higher power and
3) creates a matrix with possibility of off-diagonal feedback terms.

The author of this paper leaves it to Giannantoni to further explain his math operations. This math seems to require the solution of fractional differential equations.

A PIPELINE MODEL WITH "QUAD" HELIX INTERACTIONS

Here an attempt is made to combine two of the author's previous models (Collins, 2003, 2009) to indicate an alternate approach. The two systems are a 4-level linear pipeline model and a "quadruple" helix interaction model with differential equations:

Y_1' = $\qquad\qquad$ Y1*(1111-(Y1+Y2+Y3+Y4)) +1000-Y1
Y_2'= $\qquad$ Y1*Y2*(1111-(Y1+Y2+Y3+Y4)) +.1*Y1-Y2

$Y_3' = Y_1*Y_2*Y_3*(1111-(Y_1+Y_2+Y_3+Y_4)) +.1*Y_2-3$

$Y_4' = Y_1*Y_2*Y_3*Y_4*(1111-(Y!+Y_2+Y_3+Y_4)) +.1*Y_3-Y_4$

A check shows that an equilibrium flow solution is given by the following values: $Y_1* = 1000$, $Y_2* = 100$, $Y_3* = 10$, $Y_4* = 1$.

The transformities are $T_1 = 10$ for Y_1, $T_2 = 100$ for Y_2, $T_3 = 1000$ for Y_3 and $T_4 = 10000$ for Y_4. The feedback term is $(T_{i-1}/T_i)*Y_{i-1}$ for example $(T_2/T_3)*Y_2 = (100/1000)*Y_2 = .1*Y_2$ based on an input of 10000 to Y_1. Without the first term of each equation, the system is simply a tank-filling problem as studied in beginning differential equations.

The sum of the equilibrium flow values $Y_1*+Y_2*+Y_3*+Y_4* = 1000+100+10+1 = 1111$ is set as the resource limit for the "quad-helix" terms. It should be possible to do this process in many cases, i.e. solve for the linear equilibrium values and apply these values to limit (say 1111) the nonlinear (first) terms.

The system can be solved numerically in the *Mathematica* programming language, from initial conditions $Y_1(0) = Y_2(0) = Y_3(0) = Y_4(0) = .4$, with the result that at first the nonlinear "quad helix" terms dominate (Please see Graph 1.), so that the species Y_4 controls almost all the resources from $t = .001$ to $.01$ (Please see Graph 2.). However then the system relaxes to the linear part equilibrium values from $t = 1$ to 10 (Please see Graph 3.). It is conjectured that this happens for many ecological webs, although the *Mathematical* ordinary differential equation solver may be off even in the above case.

The speculation is that while almost all the energy is in the top transformity species (say Y_4) it may create a new species (say Y_5) by a multiplier interaction as an alternate outlet to simple relaxation to the previous equilibrium values. This would correspond to adding the differential equation:

$Y_5' = Y_1*Y_2*Y_3*Y_4*Y_5*(1111.1-(Y_1+Y_2+Y_3+Y_4+Y_5)) +.1*Y_4-Y_5$ to the system.

This process would involve
1) a "binary" operation of changing Y_4 to Y_4 plus Y_5 (increasing the dimension to 5),
2) a "duet" operation of inserting an additional power Y_5 into the interaction term of Y_5
3) a "binary-duet" feedback operation $(.1*Y_4)$ creating a matrix of higher dimension.

FRANGIBILITY AND SPECTRUM

The reader can skip this section. Actually our main goal is to describe the origin of new species that are higher in the hierarchy of tansformities, i.e. represent a new ordinality. However much of the origin of species deals with the breaking of interbreeding of a given species, which may be termed "frangibility." The above model has some good properties in this regard.

For example the equation $y' = y*(10-y) + 10-y$ can be split into $y = Y_1+Y_2$ that add back up to the y-equation:

$Y_1' = Y_1*(10-(Y_1+Y_2)) +7-Y_1$
$Y_2' = Y_2*(10-(Y_1+Y_2))+ 3-Y_2$ add to
$(Y_1+Y_2)' = (Y_1+Y_2)* (10-(Y_1+Y_2)) +10-(Y_1+Y_2)$.

Here the equilibrium values are Y1*=7 and Y2*=3, corresponding to y*=10.

For example the equations might represent a bird which divides into two species because of differences in their beaks with 70% of the total food resource being berries (for Y1) and 30% being nuts (for Y2).

This process can give rise to a spectrum of species.

For example the equation y' = y^2*(10-y) +10-y with y=Y1+Y2 can break into the two equations:

Y1'=   Y1^2*(10-(Y1+Y2)) +7 - Y1

Y2'= (2*Y1*Y2+Y2^2)*(10-(Y1+Y2)) +3 -Y2   that add to

(Y1+Y2)'=(Y1+Y2)^2*(10-(Y1+Y2)) +10 -(Y1+Y2)   compatible with the y-equation.

It might be possible to check "experimentally" if any existing species diversity follows the above type of equations.

It is also possible to work with multinomial expansions corresponding to division into Y1, Y2, and Y3, for example.

HUMAN EVOLUTION

There is a question of whether the Earth should be a machine that is progressing or a comfortable dwelling place, which goes back to at least to the work of Teilhard de Chardin and C.S. Lewis. This choice raises the question of whether there will evolve a new species of higher transformity than humans, or whether the end result will be sustainability of the status quo, or collapse of human culture. If there is a higher-transformity species, there is a question of whether it will be predatory on current-level humans, such as the Morelocks of H.G/. Wells' *Time Machine*, and present-day discussion/fiction on vampires, stealing body parts, and so on, or whether it will be some kind of noosphere development as suggested by Salthe (1996, p.223), leading to super-intelligent (or not) childlike beings.

Perhaps there will be a spectrum, based on DNA-biology gone wild.

CONCLUSION

This paper considers some questions involved in the mathematical origin of species, as mediated by computer science. There is much room for development of the models presented. The conditions of emergence of higher-ordinality species remains elusive.

ACKNOWLEDGMENTS

The author acknowledges discussion with H.T. Odum and Mr. Glenn H. Collins.

In discussing "Rationale of the Transformity Method," H.T. said something like "You have the nonlinear stuff and then at the basic level it reduces to the linear."

REFERENCES

Braunstein, Samuel, Sibasish Ghosh and Simone Severini, 2008 The laplacian of a graph as a density matrix, arXiv: quant-ph/0406165v2, p.15.

Collins, Dennis 2003 On the rationale of the transformity method, in *Emergy Synthesis 2*, ed. Mark T. Brown et al., Center for Environmental Policy, Univ. of Florida, Gainesville, FL, p.175.

Collins, Dennis 2010 On Some Continuous Empower2 Models, presented at Emergy Research Conference 6, Jan. 15, 2010, Univ. of Florida, Gainesville, FL.

Gamma, Erich, Richard Helm, Ralph Johnson and John Vlissides, 1995 *Design Patterns*, Addison-Wesley, Boston, p.97.

Giannantoni, Corrado 2009 From Transformity to Ordinality, in *Emergy Synthesis 5*, ed. By Mark T. Brown, Sharlynn Sweeney et al, Center for Envronmental Policy, Univ. of Florida, Gainesville, FL, p.581.

Odum. H.T. 1996 *Environmental Accounting*, John Wiley and Sons, NY, pp.88-109.

Salthe, Stanley N. 1996 *Development and Evolution*, MIT Press, Cambridge, MA, p.223.

Wolfe, George 2010, Thinkwell On-line Biology videos.

```
Clear All
s=NDSolve[{y1'[t]y1[t]*(1111.0-(y1[t]+y2[t]+y3[t]+y4[t]))+1000-
y1[t],y2'[t](y2[t]*y1[t])*(1111-(y1[t]+y2[t]+y3[t]+y4[t]))+.1*y1[
t]-y2[t],y3'[t](y3[t]*y2[t]*y1[t])*(1111-(y1[t]+y2[t]+y3[t]+y4[t])
)+.1*y2[t]-y3[t],
y4'[t](y1[t]*y2[t]*y3[t]*y4[t])*(1111-(y1[t]+y2[t]+y3[t]+y4[t]))
+.1*y3[t]-y4[t],y1[0].4,y2[0].4,y3[0].4,y4[0].4},{y1,y2,y3,y4},{
t,20}]
Plot[Evaluate[{y1[t],y2[t],y3[t],y4[t]}/.s],{t,0,.01},PlotRange→All]
Table[Evaluate[{y1[t],y2[t],y3[t],y4[t]}/.s],{t,.001,.01,.001}]
```

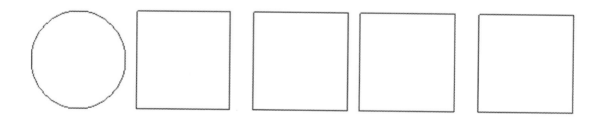

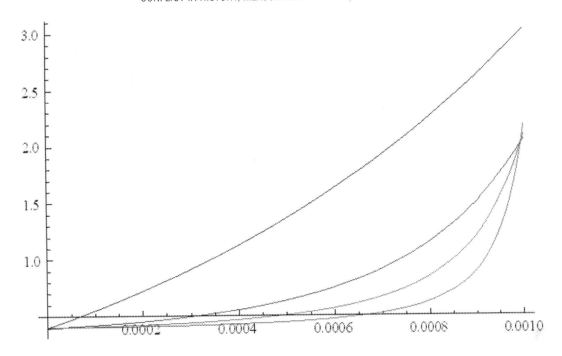

Table[Evaluate[{y1[t],y2[t],y3[t],y4[t]}/.s],{t,.0001,.0010,.0001}]
{{{0.552634,0.4216,0.408712,0.403478}},{{0.723145,0.452391,0.42147,0.408638}},{{0.913615,
0.495227,0.439895,0.416215}},{{1.12637,0.554348,0.466665,0.427486}},{{1.364,0.636171,0.5
06433,0.444809}},{{1.62939,0.750639,0.567833,0.472956}},{{1.92573,0.913579,0.668306,0.
522927}},{{2.25659,1.15096,0.84736,0.625451}},{{2.62584,1.50685,1.20958,0.898471}},{{3.03
762,2.05853,2.10016,2.18929}}}

Graph 1. Pipeline with quad helix, early evolution: Y1 top graph to Y4 at bottom. Vectors are
values of Y1,Y2,Y3,Y4 over (ten) increasing t values.

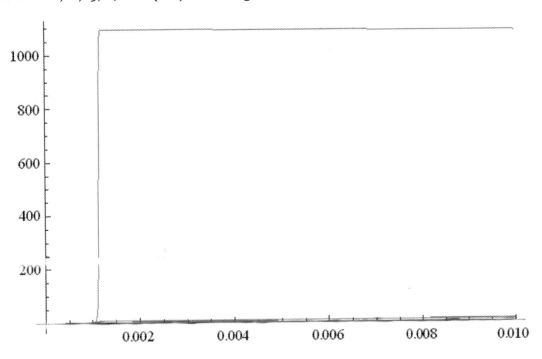

Table[Evaluate[{y1[t],y2[t],y3[t],y4[t]}/.s],{t,.001,.010,.001}]
{{{3.03762,2.05853,2.10016,2.18929}},{{4.54559,3.43182,8.30844,1094.71}},{{5.54055,3.4289,
8.30058,1093.73}},{{6.53451,3.42609,8.29273,1092.75}},{{7.52748,3.42338,8.28488,1091.76}}
,{{8.51946,3.42077,8.27704,1090.78}},{{9.51045,3.41827,8.26921,1089.8}},{{10.5005,3.41586,
8.26139,1088.82}},{{11.4895,3.41356,8.25357,1087.84}},{{12.4775,3.41136,8.24576,1086.87}}}

Graph 2. Pipeline with quad helix, middle evolution: Y4 at top, others along bottom.

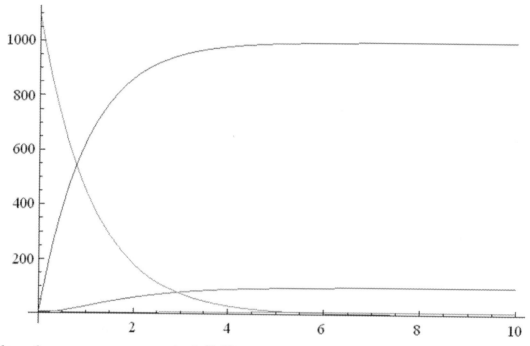

{{{633.061,27.7952,4.05993,446.084}},{{865.011,59.9536,4.54687,181.489}},{{950.34,80.313
,6.33479,74.0117}},{{981.731,90.939,7.89656,30.4331}},{{993.279,96.0012,8.9157,12.8037}},{{
997.528,98.2869,9.49009,5.69533}},{{999.091,99.2825,9.78064,2.84632}},{{999.665,99.70
48,9.91275,1.71703}},{{999.877,99.8801,9.96704,1.27588}},{{999.955,99.9518,9.98781,1.1056
2}}}

Graph 3. Pipeline with quad helix, late evolution: Y4 relaxing from about 1100 down to about 1.

Measuring the Symmetry of a Finite Group

By Dennis Glenn Collins
1519S State Rd. 119 Apt. 2
Winamac, IN 46996-8550

ABSTRACT: A method is given to measure the symmetry of any finite group according to the author's 2007 (2011) patent and examples are presented. For example the dihedral group of the square of order 8, has SYM Dih4(6) symmetry SYM = 122, and the quaternion group of order 8 has SYM symmetry = 282. The two non-isomorphic groups of order 4 have the same SYM symmetry = 7.

INTRODUCTION

Before the author's work [1], the symmetry of a geometric figure was calculated by individual cases, based on various centers and axes of symmetry and symmetry groups. However the author's work permits a calculation of symmetry for any figure and comparison of any two figures. It is natural to try to reverse the process and calculate the symmetry of any finite group, so that any two finite groups can be compared as to their amount of symmetry.

METHOD

To find the symmetry of a finite group it is necessary to measure the distance between any two group elements. However it is known that any finite group can be considered as a subgroup of a permutation group, i.e. symmetric group Sm. Further any element of a symmetric group can be presented as a product of transpositions [2], or interchanges of two symbols of the basic set for the permutation group. Since the equation ax = b can always be solved for x uniquely in a group, the distance from a to b can be considered as the length of the solution x that takes a to b, and the length of x can be counted as the minimum number of transpositions that represent x. The identity element requires no permutations and thus has length zero. A cycle with k symbols can be written as a product of k-1 transpositions [2].

According to the author's patent [3], the symmetry can be measured from the distribution of lengths over all pairs of group elements. However it is known that for any element of the group, the product of one group element with all group elements traces out the set of all group elements, perhaps in a different order. Thus if the order of the group is n, the distribution of lengths of all elements will simply be n times the distribution of lengths for any one element multiplied by all elements of the group.

A POSSIBLE GLITSCH IN THE METHOD

In the geometric figure case the distance a to b is the same as the distance b to a because of the metric structure. However a group is not necessarily abelian, so that the condition ab = ba does not necessarily hold. However if ax= b then a= b (x inverse) or b (x inverse) = a, which means the distance from b to a is measured by x inverse.

In the case of x written as a product of transpositions, x inverse can be written as a product of the same number of transpositions by reversing the permutation changes. Then the distance from a to b is the same as the distance from b to a measured as the minimum number of transpositions. Thus if the n identity elements are eliminated (as ever-present intrinsic symmetry) and the a to b and b to a distance are considered as one distance, the number of possibly-distinct distances comes out to n(n-1)/2 for a group of order n, the same as for a geometric figure with n points.

ALTERNATE METHOD

An alternate method is to consider directed distances, i.e. "vectors." In this case a group of order n would have n squared directed distances in its multiplication table and the number of pairs of equal vector distances can be calculated to determine symmetry. Here the identity (0 cycles) elements can be included on the n by n graph multiplication table. This alternate calculation will give a larger value of symmetry in general. However this method covers any multiplication table whose elements can be written as a product of cycles

EXAMPLES

First a geometric figure is calculated to illustrate the general method. Then two groups of order four, 1) C4 and 2) the "Klein Four Group" are discussed. Then 2) the five non-isomorphic groups of order eight are considered. The distribution of lengths for the n-1 non-identity elements are multiplied by n/2 because of the pairing of an element with its inverse, yielding n(n-1)/2 distances.

The author's patent [3]and further papers present three measures of symmetry, which are typically highly correlated, SYM, UPED, and PAL.

1) The two non-isomorphic groups of order four have the same symmetry. Here (4*3)/2 = 6 pairs of equal distances to be considered, which are the number of positions below the main diagonal.
   a) The cyclic group Z4 can be represented by elements e=( ), (1234), (13)(24), and (1432). Please see Figure 2. There are 4*3/2=6 pairs with d=2 having an occupation number 2*1=2 and d=3 having an occupation number 2*2 = 4.
   b) The Klein Four group can be represented by elements e= ( ), (12), (34), and (12)(34). The six pairs correspond to d= 1 having an occupation number 2*2=4 and d=2

having an occupation number 2*1 = 2. Both groups have the same symmetry, for example Sym = 2*1/2+4*3/2 = 1+6=7. The maximum possible symmetry 6*5/2=15 does not occur.

2) The five non-isomorphic groups of order eight can be taken as Dih4 = dihedral group of symmetries of the square, Z2xZ2xZ2 = symmetry group of coordinate axes flips in three dimensions, Z8 = cyclic group of rotational symmetries of an octagon, Z4xZ2, and Q = quaternion group of plus and minus i, j, k vectors in three-dimensional space.

a) Dih4(4): The symmetries of the square with 1,2,3,4 on the corners can be generated by cycles (1234) (rotations) and cycle (13) (mirror image), giving the eight elements ( ), (13), (24), (12)(34), (1234). (1432), (13)(24), (14)(23). Then the occupation numbers come out as d=1 cycle has occupation number 4* 2=8, d= 2 has occupation number 4*3=12 and d= 3 cycles has occupation number 4*2= 8, totaling 28.

b) Z2xZ2xZ2(6): The elements can be taken as generated by cycles (12),(34),(56) with elements ( ), (12), (34),(56), (12)(34), (12)(56), (34)(56), (12)(34)(56). The occupation numbers come out as d= 1 has occupation number 4*3= 12, d= 2 has occupation number 4*3= 12 and d=3 has occupation number 4*1 = 4, totaling 28.

c) Z8, the cyclic group of eight elements: The elements of addition mod 8 can be taken as 0,1,2,3,4,5,6,7. Writing the multiplication table and considering permutations, these elements correspond to cycles ( ), (01234567), (1357)(2460), (14725036), (04)(15)(26)(37), (05274163), (0642)(1753), (07654321). Then the distribution of cycles comes out as d=4 has occupation number 4*1=4, d=6 has occupation number 4*2=8 and d=7 has occupation number 4*4=16, totaling 28.

d) Z4xZ2: The elements can be taken as (0,0), (1,0),(2,0), (3,0) and (0,1), (1,1), (2,1), (3,1) with the first entry added mod 4 and the second entry added mod2. The multiplication table can be changed to numbers 1,2,3,4,5,6,7,8 corresponding to the above elements. Then by checking permutations, based on left multiplication by the given element, the corresponding cycles come out ( ), (1234)(5678), (13) (24)(57)(68), (1432)(5876), (15)(26)(37)(48), (1638)(2745), (17)(28)(35)(46), (1836)(2547). The distribution of cycles then comes out as d= 4 has occupation number 4*3=12 and d=6 has occupation number 4*4 = 16, totaling 28.

e) Q: The elements 1,i,j,k,-1,-i,-j,-k can be listed as 1,2,3,4,5,6,7,8 and the multiplication table written out in terms of numbers. By checking the permutations based on left multiplication by the given element, the corresponding cycles come out ( ), (1256)(3478), (1357)(2864), (1854)(2763), (15),(26),(37),(48), (1652)(3874), (1753) (2468), and (1854)(2763). The distribution of cycles then comes out as d=4 has occupation number 4*1=4 and d=6 has occupation number 4*6=24, totaling 28.

DISCUSSION AND BUG ANALOGY

The quaternion group Q (Sym = 282) comes out with by far the most symmetry of the groups of order eight. The reader may question why the dihedral group Dih4 of symmetries of the square, which has the least symmetry (Sym = 122), might not be considered to have more symmetry since it has rotations and mirror images. One can consider a bug analogy. If a tree trunk has many different bugs on it, it may be considered to have much diversity but not so much symmetry. If all the bugs are the same, then they all have the same shape and the symmetry is more. Also one could compare an orchard where all the trees are the same to have more symmetry than a forest with many different kinds of trees. In the quaternion case six of the eight elements of the group have the same "shape," leading to more symmetry.

Observe the three measures of symmetry Sym, Uped, and Pal are correlated with each other above 97%.

DOMAIN REDUCTION—ANOTHER HEADACHE

It may be observed that in the EXAMPLES of groups of order 8, the order of the symmetry group that generates the dihedral group is 4 versus 8, and the order of the symmetry group that generates Z2xZ2xZ2 is 6 versus 8. Thus the longest cycles available are 4 and 6 respectively versus 8. This "DOMAIN REDUCTION" changes the domain of the distribution from maximum length 8 down to a smaller number. Although intuitively this decrease might be expected to increase the symmetry SYM, in practice it seems to decrease the occupation numbers and thus the symmetry as measured by SYM. To take domain reduction into account, the size n of the group Sn applied to generate the group can be put in parentheses if it is not the same as the order of the group, i.e. the default order n of Sn is taken as the order of the group but otherwise (reduced domain) it must be included in the data.

2 a)' If the standard domain of length 8 is taken to generate the dihedral group, the cycles come out ( ), (1234)—> (1234)(5678), (13)(24)-> (13)(24)(56)(78),(1432)-> (1432)(5678), (13)-> (15)(28)(35)(48), (24)-> (16)(27)(35)(48),(12)(34)-> (17)(25)(38)(46), (14)(23)-> (18)(26)(37)(45). This changes the occupation numbers to (0,0,0,20,0,8,0) based on the domain size 8 versus 4.

2 b)' If the standard domain of length 8 is taken to generate the group Z2xZ2xZ2, the cycles come out ( ), (12)(34)(56)(78), (13)(24)(57)(68), (14)(23)(58)(67), (15)(26)(37)(48), (16)(25)(38)(47), (17)(28)(36)(43), (18)(27)(36)(45), with every non-identity element having the same occupation number 4, so that the occupation numbers come out (0,0,0, 28, 0,0,0). Thus in the standard domain, the group Z2xZ2xZ2 HAS MAXIMUM SYMMETRY.

2 d)' By checking the subgroup lattice of S6, it turns out Z4xZ2 has a cycle representation over S6 of the form ( ), (1234), (13)(24), (1432), (56), (1234)(56), (13)(24)(56), (1432)(56), so that the occupation numbers come out (4,4,12,8,0), totaling 28. As mentioned above then counter-intuitively with this change of domain Z4xZ2(6) HAS MINIMUM SYMMETRY SYM= 106 for groups of order 8.

The question arises as to which calculation should be taken as the real symmetry measure if there is both a standard and reduced domain for a given group. This is similar to an optical illusion, where there are two different ways to look at a given geometric figure. For example in the case of Z2xZ2xZ2, looked at as axis flips in three-dimensional space Z2xZ2xZ2(6) there is not that much symmetry SYM = 138. Looked at as addition of triples mod2, there is much more symmetry with Z2xZ2xZ2 having SYM = 28*27/2 = 378. What is needed is what is called a regularization method, one of which is to apply the standard domain and cycle representation.

A COUPLE SIMPLE THEOREMS

The maximum symmetry occurs if only one non-zero occupation number occurs, that is if the distribution is concentrated on one value, such as 4 transpositons above 2 b)' for Z2xZ2xZ2. As can be calculated via the patent, this case corresponds to minimum entropy ENT since there is no uncertainty about the cycle structure. An interesting question is if there is any relation between the ENT entropy and other definitions of group entropy.

Theorem 1. The group $(Z_2)^{\wedge}n$ has maximum symmetry SYM on the standard domain.

Proof: Every element x has the property $x*x = e =$ identity, so that it has n transpositions in the cycle representation, so that only the occupation number $n(n-1)/2$ occurs at value $n/2$. This pattern corresponds to a Cayley graph of "all spokes."

Theorem 2. A (cyclic) group of prime order p has maximum symmetry SYM on the standard domain.

Proof: Every element has a cycle representation of length p-1, so that the only occupation number is p-1 with height $p(p-1)/2$. This pattern corresponds to a Cayley graph that is "all wheel rim" with no subgroups.

CONCLUSIONS

The method provides a reasonable way to measure the symmetry of a finite group, which depends only on knowing the cycle structure of the group, which can be obtained via *Mathematica* in some cases. There are many open math questions, such as finding the maximum symmetry for a given order n of a group, and the continuous version. There is the question if a given group can have two different cycle representations over the same reduced domain, leading to even more ambiguity.

REFERENCES

[1] Collins, Dennis G. 2007 Algorithm to Measure Symmetry and Positional Entropy of n Points, *General Systems Bulletin*, Vol. XXXVI, pp.15-20.

[2] McCoy, Neal H 1972., *Fundamentals of Abstract Algebra*, Allyn and Bacon, Inc., Boston, pp.165-166.

[3] Collins, Dennis G. Jan. 18, 2011 U.S. Patent # 7,873,220 B1 ALGORITHM TO MEASURE SYMMETRY AND POSITIONAL ENTROPY OF A DATA SET.

[4] "List of Small Groups" Wikipedia http://en.wikipedia.org/wiki/List_of_small_groups

[5] "The Subgroups of S6," http://schmidt.nuigalway.ie/subgroups

```
Q1={{1,i,j,k,-1,-i,-j,-k},{i,-1,k,-j,-i,1,-k,j},{j,-k,-1,i,-j,k,1,-i},{k,j,-i,-1,-k,-j,i,1},{-1,-i,-j,-k,1,i,j,k},{-i,1,-k,j,i,-
1,k,-j},{-j,k,1,-i,j,-k,-1,i},{-k,-j,i,1,k,j,-i,-1}};
NC={{0,6,6,6,4,6,6,6},{6,4,6,6,6,0,6,6},{6,6,4,6,6,6,0,6},{6,6,6,4,6,6,6,0},{4,6,6,6,0,6,6,6},{
6,0,6,6,6,4,6,6},{6,6,0,6,6,6,4,6},{6,6,6,0,6,6,6,4}};
MatrixForm[Q1]
MatrixForm[NC]
({
{1, i, j, k,-1,-i,-j,-k},
{i,-1, k,-j,-i, 1,-k, j},
{j,-k,-1, i,-j, k, 1,-i},
{k, j,-i,-1,-k,-j, i, 1},
{-1,-i,-j,-k, 1, i, j, k},
{-i, 1,-k, j, i,-1, k,-j},
{-j, k, 1,-i, j,-k,-1, i},
{-k,-j, i, 1, k, j,-i,-1}
})
({
{0, 6, 6, 6, 4, 6, 6, 6},
{6, 4, 6, 6, 6, 0, 6, 6},
{6, 6, 4, 6, 6, 6, 0, 6},
{6, 6, 6, 4, 6, 6, 6, 0},
{4, 6, 6, 6, 0, 6, 6, 6},
{6, 0, 6, 6, 6, 4, 6, 6},
{6, 6, 0, 6, 6, 6, 4, 6},
{6, 6, 6, 0, 6, 6, 6, 4}
})


Clear All
Dih4={8,12,8,0,0,0,0};
Z2xZ2xZ2={12,12,4,0,0,0,0};
Z8={0,0,0,4,0,8,16};
Z4xZ2={0,0,0,12,0,16,0};
Q={0,0,0,4,0,24,0};
L1=ListPlot[Dih4,PlotMarkers→{o},Filling->Axis,PlotRange→28];
L2=ListPlot[Z2xZ2xZ2,PlotMarkers→{o},Filling→Axis,PlotRange→28];
L3=ListPlot[Z8,PlotMarkers→{o},Filling→Axis,PlotRange→28];
L4=ListPlot[Z4xZ2,PlotMarkers→{o},Filling→Axis,PlotRange→28];
L5=ListPlot[Q,PlotMarkers→{o},Filling→Axis,PlotRange→28];
G8={Dih4,Z2xZ2xZ2,Z8,Z4xZ2,Q}
Sym=Table[Sum[G8[[i,j]]*(G8[[i,j]]-1)/2,{j,1,7}],{i,1,5}]
Uped=Table[Sum[If[G8[[i,j]]0,0,N[G8[[i,j]]*Log[G8[[i,j]]]]],{j,1,7}],{i,1,5}]
Pal=Table[Sum[N[G8[[i,j]]*Exp[1-G8[[i,j]]/28]],{j,1,7}],{i,1,5}]
GraphicsGrid[{{"Dih4","Z2xZ2xZ2","Z8","Z4xZ2","Q"},{L1,L2,L3,L4,L5},Table[Sym[[k]],{k,1,
5}],Table[Uped[[m]],{m,1,5}],Table[Pal[[p]],{p,1,5}]}]
```

All Clear
{{8,12,8,0,0,0,0},{12,12,4,0,0,0,0},{0,0,0,4,0,8,16},{0,0,0,12,0,16,0},{0,0,0,4,0,24,0}}
{122,138,154,186,282}
{63.0899,65.1829,66.5421,74.1803,81.8185}
{53.9332,51.9248,50.3285,45.8105,37.1112}

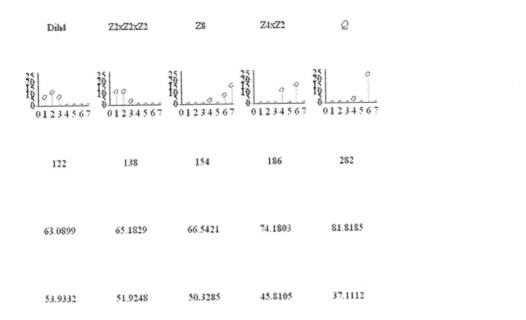

Dih4	Z2xZ2xZ2	Z8	Z4xZ2	Q
122	138	154	186	282
63.0899	65.1829	66.5421	74.1803	81.8185
53.9332	51.9248	50.3285	45.8105	37.1112

Correlation[Sym,Uped]
0.977799
Correlation[Sym,Pal]
-0.996263
Correlation[Uped,Pal]
-0.99172

Clear All
Dih4={8,12,8,0,0,0,0};
K4={4,2,0};
Z4={0,2,4};
L1=ListPlot[K4,PlotMarkers→{o},Filling→Axis,PlotRange→6];
L2=ListPlot[Z4,PlotMarkers→{o},Filling→Axis,PlotRange→6];
G8={K4,Z4}
Sym=Table[Sum[G8[[i,j]]*(G8[[i,j]]-1)/2,{j,1,3}],{i,1,2}]
Uped=Table[Sum[If[G8[[i,j]]0.0,N[G8[[i,j]]*Log[G8[[i,j]]]]]],{j,1,3}],{i,1,2}]
Pal=Table[Sum[N[G8[[i,j]]*Exp[1-G8[[i,j]]/6]],{j,1,3}],{i,1,2}]
GraphicsGrid[{{"Klein4","Z4 Cyclic"},{L1,L2},Table[Sym[[k]],{k,1,2}],Table[Uped[[m]],{m,1,2}],Table[Pal[[p]],{p,1,2}]}]

All Clear
{{4,2,0},{0,2,4}}
{7,7}
{6.93147,6.93147}
{9.47792,9.47792}

Klein4   Z4 Cyclic

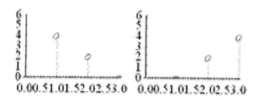

7   7

6.93147   6.93147

9.47792   9.47792

```
01    !LEFT RIGHT TOUGH TENDER  ARV 20 JULY 2007
02    !COPY FROM APPLE TO IBM
03    !CONVERT QBASIC TO TRUEBASIC
04    !MAKE G=1 FOR TIME GRAPH AND G=2 FOR CHOICE V.S. (LEFT-10*RIGHT)
05    !TOUGHTENDER H.T. ODUM AND DAVID SCIENCEMAN (APPLE)
06    OPTION NOLET
07    CLEAR
08    Set Color "Black"
09    Set Window -30,279,-30,279
10    Box Lines 0,279,0,180
12    G=21
14    IF G=2 THEN GOTO 20
20    J1=20
25    J3=30
30    K0=1.9
35    E1=0.5
37    E2=0.1
40    K1=0.029
50    K2=0.005
60    K3=0.1
70    K4=0.1
80    K5=0.02
90    K6=0.5
95    K7=0.05
100   K8=0.026
105   K9=0.03
110   L1=0.002
115   L2=0.00005
120   L3=0.01
121   L4=0.0005
130   L5=0.005
131   L6=0.002
132   L7=0.01
133   TH=100
135   T=1
137   Z0=1
140   T0=1
142   C0=0.3
144   L0=2
146   R0=1
148   Q0=10
150   I=0.5
155   C=5
160   L=50
170   Q=100
180   R=5
200   J2=J1/(1+K0*C)
205   J4=J3/(1+K1*L*R)
207   LR=Abs(L-10*R)
208   IF LR>10 THEN LR=10
210   DL=K8*Q-K9*L-L1*J4*L*R +L3*W*Q
220   DQ=K1*E1*J4*L*R +K6*J2*C -K7*Q -K8*Q -L6*R*Q -L5*Q -L7*Q*(10-LR)
225   DC=K2*Q*(10-LR) -K3*J2*C -K4*C
230   DR=Y*L4*Q+E2*L6*R*Q-L2*R*L*J4-K5*R
231   A=A+1
232   IF A<10 THEN GOTO 250
233   IF Q<TH THEN Y=Y+1
234   IF Q<TH THEN A=1
235   IF Y>1 THEN Y=0
238   IF Y=1 THEN W=0
```

```
239   IF Y=0 THEN W=1
250   L=L+DL*I
255   IF L<0.1 THEN L=0.1
260   Q=Q+DQ*I
265   IF Q<0.1 THEN Q=0.1
270   C=C+DC*I
275   IF C<.1 THEN C=0.1
280   R=R+DR*I
285   IF R<0.1 THEN R=0.1
290   Z=K0*J2*C +K1*J4*L*R
291   Set Color "Black"
297   IF G=1 THEN GOTO 310
298   Set Color "Red"
300   PLOT 100+(5*LR),90-C/C0
305   IF G=2 THEN GOTO 340
306   Set Color "Green"
310   PLOT T/T0,100+L/L0
313   Set Color "Cyan"
314   PLOT T/T0,30+Q/Q0
315   Set Color "Blue"
330   PLOT T/T0,100+R/R0
331   Set Color "Black"
333   PLOT T/T0,20+C/C0
334   Set Color "Red"
336   PLOT T/T0,100-Z/Z0
340   T=T+I
400   IF T/T0< 279 THEN GOTO 200
500   END
```

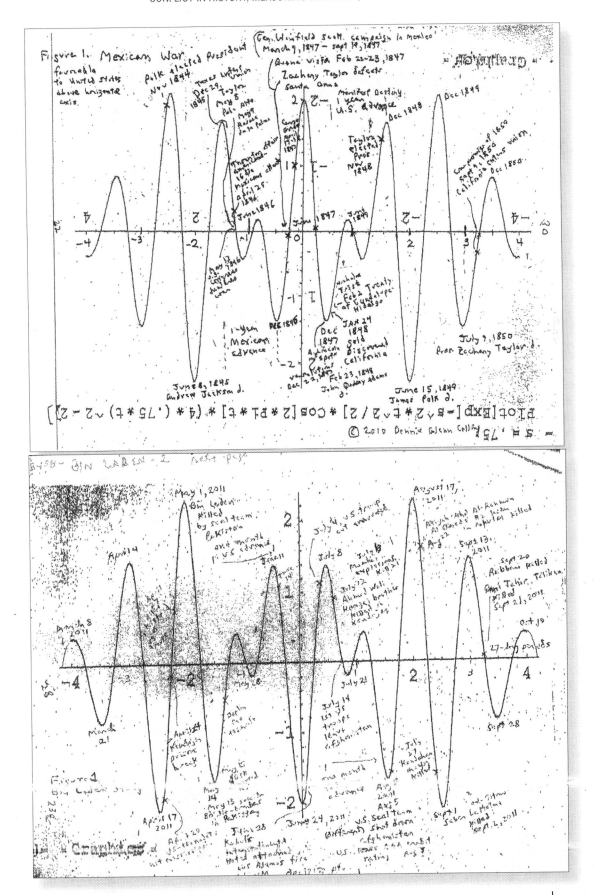

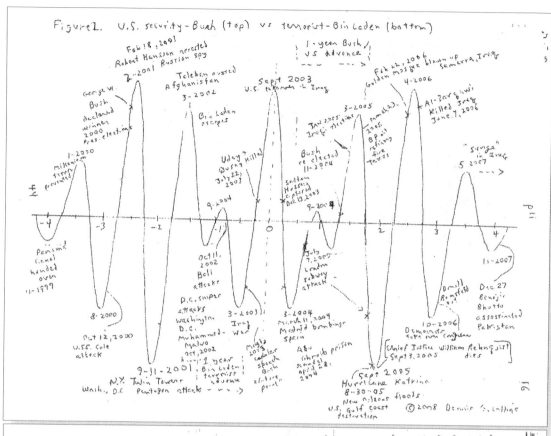

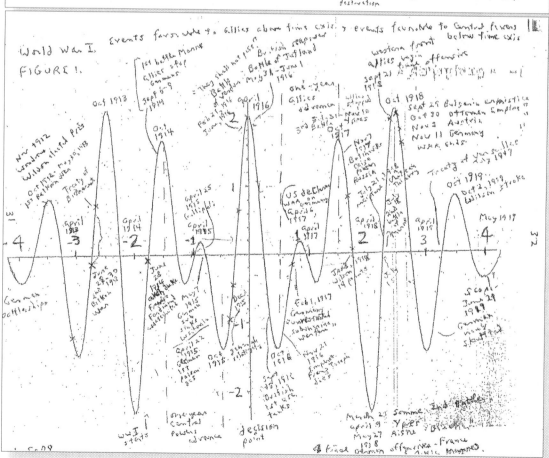

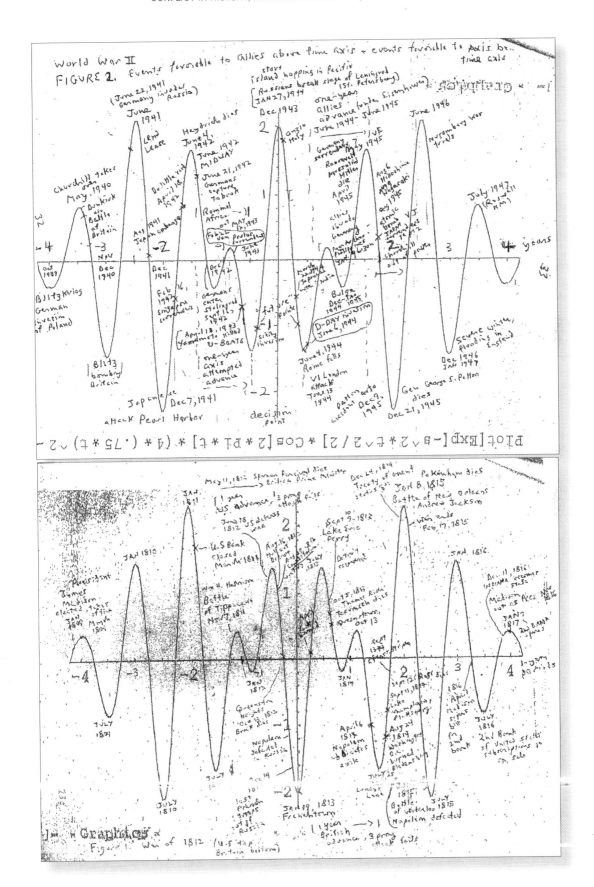

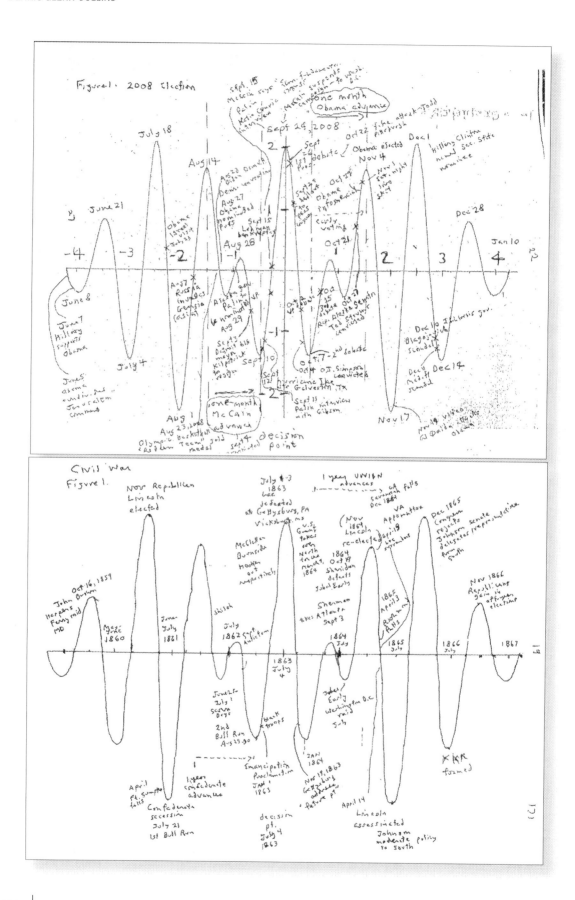

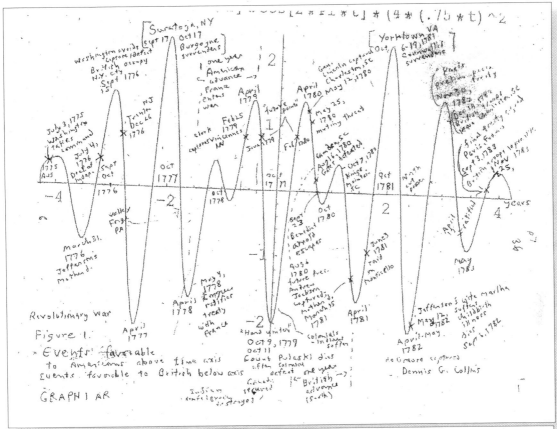

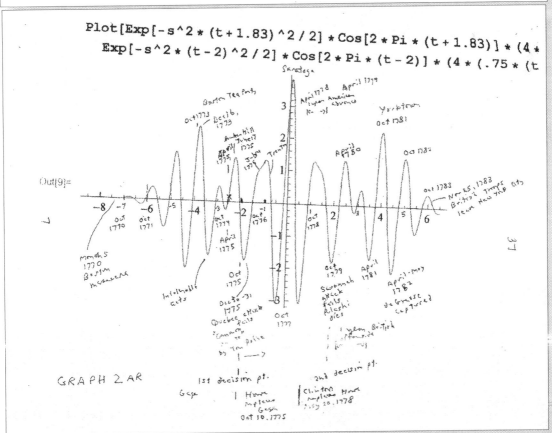

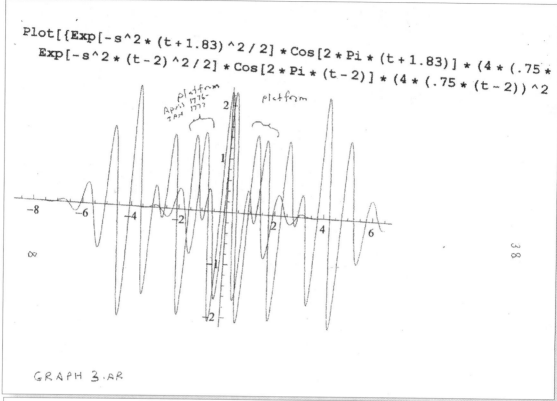

```
Plot[{Exp[-s^2 * (t+1.83)^2/2] * Cos[2 * Pi * (t+1.83)] * (4 * (.75 *
    Exp[-s^2 * (t-2)^2/2] * Cos[2 * Pi * (t-2)] * (4 * (.75 * (t-2))^2
```

GRAPH 3.AR

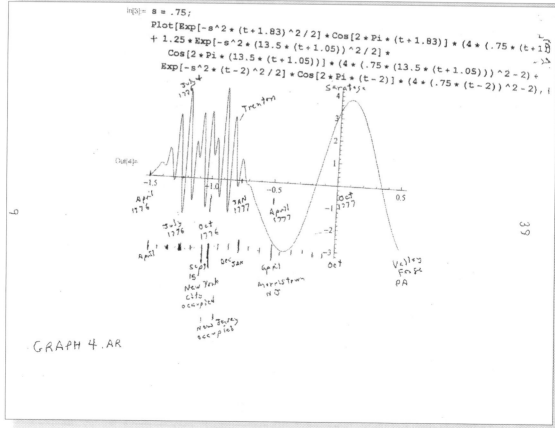

```
In[3]:= s = .75;
    Plot[Exp[-s^2 * (t+1.83)^2/2] * Cos[2 * Pi * (t+1.83)] * (4 * (.75 * (t+1.0
    + 1.25 * Exp[-s^2 * (13.5 * (t+1.05))^2/2] *
        Cos[2 * Pi * (13.5 * (t+1.05))] * (4 * (.75 * (t+1.05))^2 /
    Exp[-s^2 * (t-2)^2/2] * Cos[2 * Pi * (t-2)] * (4 * (.75 * (t-2))^2 - 2),
```

GRAPH 4.AR

136

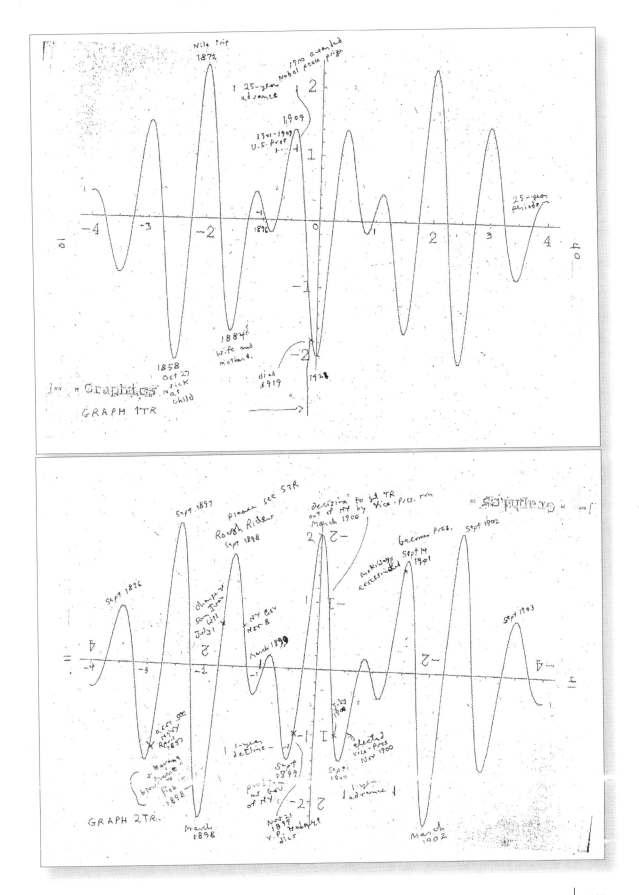

```
(4 * (.75 * (12 * (t + 1.5)))^2 - 2), -Exp[-s^2 * (12 * (t + 1.5))] *
Cos[2 * Pi * (t - 3)] * (4 * (.75 * (t - 3))^2 - 2)}, {t, -6.5, 8.5}, Pl
```

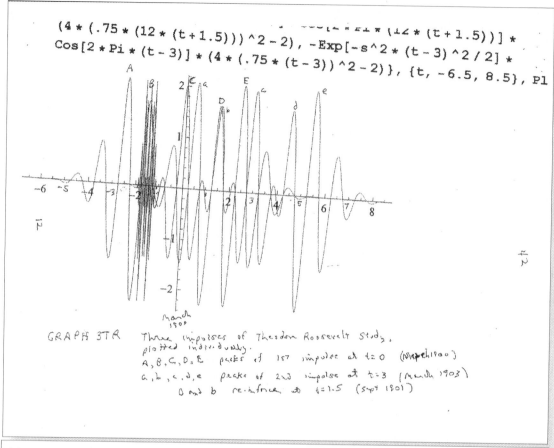

GRAPH 3TR   Three impulses of Theodore Roosevelt study,
plotted individually.
A, B, C, D, E   peaks of 1st impulse at t=0 (March 1900)
a, b, c, d, e   peaks of 2nd impulse at t=3 (March 1903)
B and b reinforce at t=1.5 (Sept 1901)

```
In[1]:= s = .75;
Plot[{-Exp[-s^2 * (t + 0)^2 / 2] * Cos[2 * Pi * (t + 0)] * (4 * (.75 *
    Exp[-s^2 (12 * (t + 1.5))^2 / 2] * Cos[2 * Pi * (12 * (t + 1.5))]
    (4 * (.75 * (12 * (t + 1.5)))^2 - 2) - Exp[-s^2 * (t - 3)^2 / 2] *
    Cos[2 * Pi * (t - 3)] * (4 * (.75 * (t - 3))^2 - 2)}, {t, -6.5, 8.
```

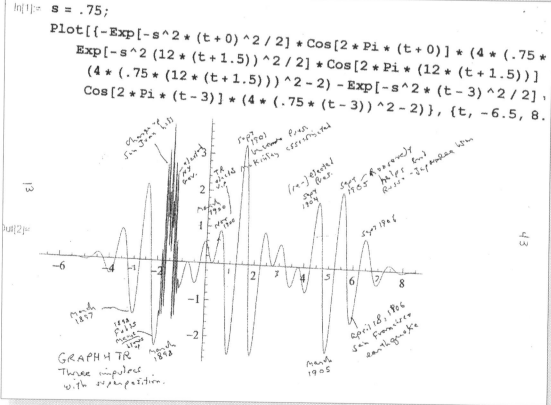

GRAPH 4 TR
Three impulses
with superposition.

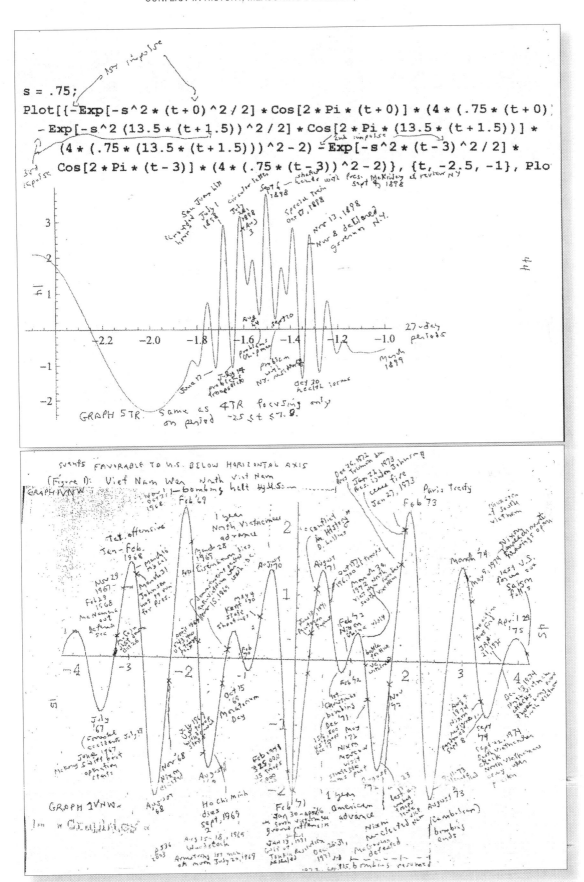

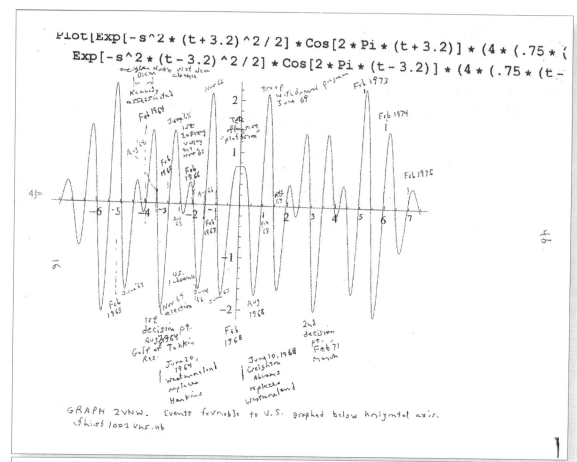

GRAPH 2VNW. Events favorable to U.S. graphed below horizontal axis.
(f.hist 1002 vns.nb)

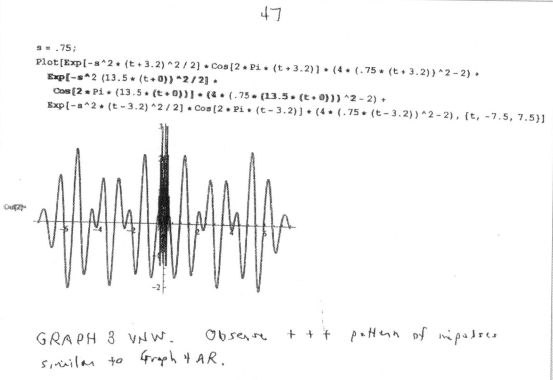

GRAPH 3 VNW. Observe + + + pattern of impulses
similar to Graph 4 AR.

48

```
s = .75;
Plot[Exp[-s^2 * (t + 3.2)^2 / 2] * Cos[2 * Pi * (t + 3.2)] * (4 * (.75 * (t + 3.2))^2 - 2) +
  Exp[-s^2 (13.5 * (t + 0))^2/2] *
  Cos[2 * Pi * (13.5 * (t + 0))] * (4 * (.75 * (13.5 * (t + 0)))^2 - 2) +
  Exp[-s^2 * (t - 3.2)^2 / 2] * Cos[2 * Pi * (t - 3.2)] * (4 * (.75 * (t - 3.2))^2 - 2), {t, -1.5, 1.5}]
```

GRAPH 4 VNW.

1967   1968

periods favorable to U.S. graphed below horizontal axis

Nov 24, 1967   2½ periods before JAN 31 (Tet) 1968   American win at Dak To.

Dec 8, 1967   2 periods before JAN 31   North Viet Nam planning

Dec 20, 1967   American reinforce Khe Sanh

Dec 20 to JAN 16, 1968   one period North Viet Namese advance
1967   in assembling Tet attack force versus
        Gen. Westmoreland claims of near end of war

JAN 21, 1968   ½ period   attack on Khe Sanh begins

JAN 31, 1968   0 = impulse decision point of 27-day periods
        with some advance preparation, Americans counter Tet offensive
        although some defeats at 1-day period

Feb 13, 1968   ½ period   Most "successful" Tet offensive at Hue
        leads to massacre of 3000 civilians by
        North Viet Nam

Feb 13 - March 12, 1968   one period of advance by American forces;
        (cf photo of shooting suspected North vietnamese agent);
        however leads to MyLai (March 16, 1968) as city
        fighting makes it difficult to tell enemy

March 26, 1968   2 periods after JAN 31, 1968   Johnson decision
        not to run (March 31) for re-election leads to 'cut-and-run'
        U.S. policy, major North Viet Nam victory (political) of Tet offensive

April 8, 1968   2½ periods after JAN 31, 1968
        Americans relieve Khe Sanh siege, ending possibility
        of another Dien Bien Phu at Khe Sanh

Observe "platform" makes North Vietnamese defeats not as serious
as they would otherwise be.

18

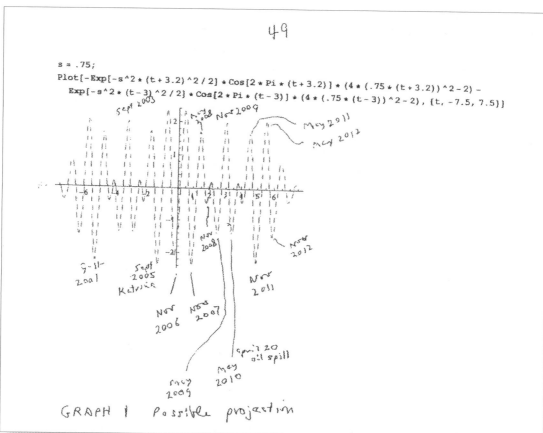

49

```
s = .75;
Plot[-Exp[-s^2 * (t + 3.2)^2 / 2] * Cos[2 * Pi * (t + 3.2)] * (4 * (.75 * (t + 3.2))^2 - 2) -
    Exp[-s^2 * (t - 3)^2 / 2] * Cos[2 * Pi * (t - 3)] * (4 * (.75 * (t - 3))^2 - 2), {t, -7.5, 7.5}]
```

GRAPH 1   Possible projection

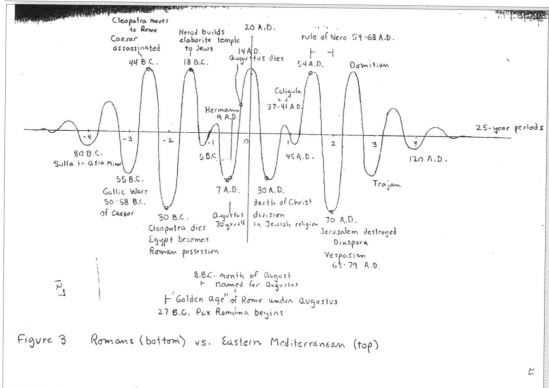

Figure 3   Romans (bottom) vs. Eastern Mediterranean (top)

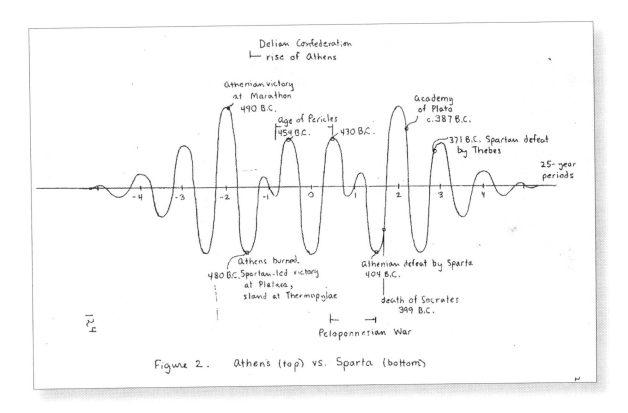

Figure 2. Athens (top) vs. Sparta (bottom)

INDEX

YOUR BOOK

From: **david scienceman** (davidmitchell.au@hotmail.com)
Sent: Wednesday, October 19, 2011 5:13:07 PM
To: Dennis Collins (d_collins_pr@hotmail.com)

Dear Professor Dennis Collins.

1. Permission granted to mention me in your POLITICAL SPECTRUM MODELS book. HOWEVER, please use my Bathurst address - I don't want strangers bashing on the door here in Castle Hill. It is: Dr. David M. Scienceman, The Farm Club of Australia, P.O. Box 307, Bathurst, NSW, 2795, AUSTRALIA. Email: davidmitchell.au@hotmail.com

ALSO, please insert somewhere the LEFT-RIGHT-TOUGH-TENDER program and diagram which Odum and Florence and I produced over so many years, I finally got it working perfectly last year. Otherwise your programs are unintelligble. I did not try and consult Dr Glenn Wilson because I did not want to interfere with the peer review process.

2 Dr Mark Brown usually quotes me very wrongly. The often quoted article by Brown and Herendeen in Ecological Economics (1996), (pages 219-235). always horrified me, but I had nothing to do with it and I can do nothing about it. There is a vast difference between 'available energies' and 'energy memories'. I was present in North Carolina with Odum when it was started. EMERGY is not equal to the Available Energy used up It is the amplifier effect PROPORTIONAL to the available energy previously required to produce a product.

3. This computer will not accept EXTEND programs and must be replaced soon. We do not have or want anything to do with FACEBOOK or TWITTER etcetera. Dr Dand Campbell sems to have got into difficulties in Swansea, Wales, after having his passport and credit cards stolen according to a very suspicious email here. Florence advises me hot to touch it.

4. My RELISCIENCE 'article' was only SCRUFFY NOTES. I have restricted myself to drawing the BASIC CHRISTIANITY STORY and its relation to the OSIRIS STORY. The book by Karen Armstrong (2009) (The CASE for GOD , is full of very valuable references even though she seems totally ignorant of ENERGETICS let alone EMERGY.

5 For safety, I am always very formal in my correspondences. Yours Sincerely, David M. Scienceman.